Dunnock Behaviour and Social Evolution

Oxford Series in Ecology and Evolution
Edited by Robert M. May and Paul H. Harvey

The Comparative Method in Evolutionary Biology
Paul H. Harvey and Mark D. Pagel

The Causes of Molecular Evolution
John H. Gillespie

Dunnock Behaviour and Social Evolution
N.B. Davies

Natural Selection: Domains, Levels,
 and Challenges
George C. Williams

Dunnock Behaviour and Social Evolution

N.B. DAVIES

Department of Zoology, University of Cambridge
Fellow of Pembroke College

with drawings by
DAVID QUINN

Oxford   New York   Tokyo
OXFORD UNIVERSITY PRESS
1992

Oxford University Press, Walton Street, Oxford OX2 6DP
Oxford New York Toronto
Delhi Bombay Calcutta Madras Karachi
Petaling Jaya Singapore Hong Kong Tokyo
Nairobi Dar es Salaam Cape Town
Melbourne Auckland
and associated companies in
Berlin Ibadan

Oxford is a trade mark of Oxford University Press

Published in the United States
by Oxford University Press, New York

A catalogue record for this book is available from the British Library

Library of Congress Cataloging in Publication Data
Davies, N. B. (Nicholas B.), 1952-
Dunnocks and breeding systems / N.B. Davies ; with drawings by David Quinn.
(Oxford series in ecology and evolution)
1. Dunnock—Behavior. 2. Sexual behavior in animals. I. Title. II. Series.
696.P266D38 1992   598.8'73—dc20   92-4302

ISBN 0 19 854674 2 (h/b)
0 19 854675 0 (p/b)

Set by Advance Typesetting Limited
Printed in Great Britain by
Biddles Ltd, Guildford & King's Lynn

*For my father
Anthony Barry Davies,
and in memory of my mother
Joyce Margaret Davies*

Acknowledgements

My first, and greatest, thanks are to my parents who have encouraged my interest in natural history ever since, at the age of six, I announced that I was going to be a bird watcher when I grew up. Unlike the parents of a colleague, they have never asked when I am going to get a proper job! To them I dedicate this book. My wife Jan, and daughters Hannah and Alice, have provided great support and have helped in numerous ways, from tolerating the early morning alarm clock to painting model eggs. I did not realise how much my life has been dominated by a little brown bird until I saw how Hannah had filled in a school form, entering her father's occupation as 'dunnocks'.

The Cambridge University Botanic Garden has proved a wonderful study site, a nature reserve within a city and just a short cycle ride from both home and the Zoology Department; I thank the two directors during the ten years of the study, Max Walters and Donald Pigott, together with the Superintendent Peter Orriss, the Supervisor Norman Villis, and all the gardening staff, for allowing me the freedom of the Garden and for their help and friendship. The work has been funded at various stages by grants from the Natural Environment Research Council, the Science and Engineering Research Council and the Royal Society, for which I am most grateful.

Many people have helped with the field work and ideas and much of the book is based on joint publications, listed in the references, and acknowledged throughout the text. Special thanks must go to Ben Hatchwell, whose appearances in the book are almost as frequent as the dunnocks themselves. The field work we did together from 1988 to 1990 was especially productive and enjoyable.

Others, whose help is acknowledged with gratitude include the following: David Harper for helping to colour-ring dunnocks during the first three years, when he was studying robins in the Garden; Arne Lundberg, from Uppsala University, who visited for two summers, financed by a grant from the Swedish Natural Science Research Council, and collaborated with an experiment described in Chapter 4; Philip Byle, who studied the dunnocks for his PhD from 1985 to 1987, and kept the population colour-ringed while I went off to study cuckoos; Alasdair Houston, who has provided theoretical insights and has collaborated with work described in Chapters 9 and 10; Terry Burke, Mike Bruford and Tim Robson, from Leicester University, who did the DNA fingerprinting to elucidate maternity and paternity, described in Chapters 7 and 12; Tim Birkhead who analysed the structure of dunnock sperm storage tubules, described in Chapter 6; Haven Wiley, who visited for a summer, on sabbatical leave from the University of North Carolina, and did some song play-back experiments, described in Chapter 6; Walter Koenig and Janis Dickinson, visitors from the University of California, Berkeley, who provided a gentle reminder that there is a species, the acorn woodpecker, which has an even more complex mating system and social organisation than the dunnock; André Desrochers, Robert Magrath and Lisa Petit, who helped to ring dunnocks and find nests, and provided great companionship during their studies of blackbirds and linnets in the Garden; Michael Brooke, who collaborated with the cuckoo work described in Chapter 13; Jari Tuomenpuro, who visited one winter from the University of Helsinki, and whose study of dunnocks in Finland has provided useful comparisons.

Others who have helped with ideas through discussion or correspondence include Patrick Bateson, Jerram and Esther Brown, Mike Birkhead, Tim Clutton-Brock, Mark Elgar, David Gibbons, David Hull, Yoshiyuki Matuzaki, Masahiko Nakamura, Josephine Pemberton, Michael Ruse, Barbara and David Snow, Duncan Wood and Satoshi Yamagishi. I acknowledge in particular the inspiration provided by John Krebs and Chris Perrins, mentors since my days as a research student at the Edward Grey Institute, Oxford.

I am grateful to Gabriel Horn, Professor of Zoology at Cambridge, who granted me sabbatical leave so I could write this book, and to the Master, Lord Adrian, and the Fellows of Pembroke College, who allowed me leave from College duties. John Andrews and Joy Schreiber have been a great help with administrative matters in the Zoology Department and I thank Neal Maskell, Frances Wei and John Rodford for their help with preparing the figures. The inclusion of some photographs by the late W.B. Carr is by the kind permission of Alan and Margery Wilkins. In my old-fashioned way, I still prefer to write everything out in pen and ink and I am especially grateful to Ann Jeffrey for the skill and extraordinary speed with which she has converted this into a typescript, showing patience and good humour throughout. I thank Robert May and Paul Harvey for their enthusiasm and editorial help, and for agreeing to publish this book in their series.

David Quinn has done me a great honour in providing the illustrations for the book. His beautiful drawings capture the 'jizz' of the dunnocks perfectly and not only help to convey the interest of their behaviour but also act as a reminder that real and wonderful animals lie behind all the simplistic graphs and tables.

The book includes new analyses of the long-term data from the ten years of the study (especially Chapters 3, 4 and 5) together with a summary of work previously published in various papers. In many cases the significance of the earlier results was not realised at the time and my interpretation or emphasis has changed because of larger sample sizes or other findings, so I hope that those who are familiar with some of the published papers will still find new ideas of interest. During the study, the answers to the various problems often emerged out of sequence. For example, the initial calculations of reproductive success (Davies and Houston 1986) were made in ignorance of paternity and maternity, which it was not possible to measure until the advent of DNA fingerprinting (Burke *et al.* 1989). The interpretation of the dunnock's strange pre-copulatory display (Davies 1983) was made in ignorance of the structure of the female's sperm storage tubules (T.R. Birkhead *et al.* 1991), and so on. I have told the story in the most logical sequence, not in the sequence in which the discoveries were made. I hope this will make the results more interesting and it easier for the reader to see where the unsolved puzzles lie. Perhaps the book will encourage someone to try and solve them.

Cambridge   N.B.D.
October 1991.

Contents

1 Why dunnocks? 1

1.1 Beginning with the Reverend Morris 1
1.2 Species versus problem-oriented studies 4
1.3 Studying species-typical behaviour versus individual differences 6
1.4 Conflicts of interest 7
1.5 Summary 8

2 Study species and study area 10

2.1 The accentors 10
2.2 The dunnock 10
2.3 The study area 15
2.4 A colour-ringed population 17
2.5 The advantages of field experiments 19
2.6 Arrangement of the book 21
2.7 Summary 22

3 Population structure and the variable mating system 24

3.1 Territory overlap of males and females 24
3.2 Frequencies of the different mating systems 27
3.3 Population structure 28
3.4 Breeding season mortality 33
3.5 Winter mortality 34
3.6 Sex ratio and variation in mating systems between years 36
3.7 Origins of individuals in the same mating system 37
3.8 Comparison with other dunnock studies 41
3.9 Summary 41

4 Territorial behaviour: competition for habitat and mates 46

4.1 Influence of male and female territory size on mating system 46
4.2 How the various mating combinations arise 48
4.3 Mating system changes due to mortality: natural experiments 53
4.4 Factors influencing female territory size 56
4.5 Experimental changes caused by feeders 60

4.6   Removal experiments to investigate male defence of females   67
4.7   Summary   71

5   Factors influencing an individual's competitive success   74

5.1   Body size and age   74
5.2   Body size and mating system   74
5.3   Age and mating system   76
5.4   Male status and territory value: remove and release experiments 81
5.5   Summary   85

6   Mate guarding and mating: sexual conflict   87

6.1   Breeding conflict begins   87
6.2   The breeding cycle   87
6.3   Mate guarding behaviour   88
6.4   Intensity of mate guarding   91
6.5   Duration of the mate guarding period   96
6.6   The value of mate guarding: removal experiments   99
6.7   Female responses to individual male songs   100
6.8   Copulations   103
6.9   An extraordinary pre-copulation display: cloaca pecking   106
6.10 Sperm competition and the reproductive organs of males
      and females   109
6.11 Summary   114

7   Relating behaviour to maternity and paternity   117

7.1   Male mating success and parental care   117
7.2   Does male mating access reflect paternity?   119
7.3   DNA fingerprinting   120
7.4   Mating behaviour and paternity   123
7.5   Paternity and parental care   126
7.6   A puzzle—why no paternity marker?   128
7.7   Summary   130

8   Reproductive output from the different mating systems   132

8.1   Introduction   132
8.2   Provisioning of nestlings   132
8.3   Provisioning rate and nestling weight   134
8.4   Nestling weight and survival   136
8.5   Number of young fledged as a measure of reproductive success 138
8.6   Failed breeding attempts: interference and infanticide   140
8.7   Successful breeding attempts: parental care and fledging success 144
8.8   Removal experiments and matched comparisons   146

8.9 Variation in clutch size in relation to expected male help   149
8.10 Summary   151

9 Individual reproductive success in the various mating systems: conflicts of interest   154

9.1 Why measure individual reproductive success?   154
9.2 Lifetime success or short-term measures?   154
9.3 Success per attempt or per season?   155
9.4 Calculation of seasonal reproductive success   159
9.5 Do individual reproductive payoffs make sense of behavioural conflicts?   161
9.6 Polygynandry: behaviour and reproductive success   165
9.7 Sexual conflict and the variable mating system   166
9.8 Summary   168

10 Parental effort by males and females in pairs and trios   169

10.1 The evolution of stable cooperation   169
10.2 Provisioning of nestlings by pairs and trios   171
10.3 Reactions to changes in effort by others   173
10.4 Care of fledglings: brood division   177
10.5 Summary   178

11 How males allocate effort between broods in polygyny and polygynandry   180

11.1 The choices facing males   180
11.2 The value of male parental care   180
11.3 Reproductive allocation by males in polygyny   181
11.4 Reproductive allocation by males in polygynandry   184
11.5 How best to allocate care between broods   192
11.6 Summary   193

12 Paternity and parental effort: how good are male chick feeding rules?   195

12.1 Do dunnocks have to be clever to behave adaptively?   195
12.2 Influence of natural variation in mating access on paternity and chick feeding   196
12.3 Influence of experimental variation in mating access on chick feeding   199
12.4 Do males use the onset of laying to value their copulations? An experiment with model eggs   201
12.5 Are male responses adaptive? Influence of removals on paternity   202

12.6 Does share of matings determine parental effort? 205
12.7 Why does paternity loss influence male work rate in trios but not in pairs? 209
12.8 How females might allocate matings to maximise male help in trios 210
12.9 Summary 212

13 Parasitism by cuckoos 214

13.1 Introduction 214
13.2 The curious habits of the cuckoo 214
13.3 Cuckoo hosts and host-egg mimicry 216
13.4 Experiments with model eggs to test host discrimination 218
13.5 Why, if dunnocks do not discriminate, do dunnock-cuckoos lay distinctive eggs? 224
13.6 Why do dunnocks not discriminate cuckoo eggs? 227
13.7 Are dunnocks recent hosts? 230
13.8 Why do dunnocks not discriminate cuckoo chicks? 232
13.9 Summary 234

14 Beyond dunnocks: sexual conflict, parental care and mating systems 235

14.1 Three main conclusions 235
14.2 Sexual conflict and mating systems 235
14.3 From conflict in polyandry to cooperation 238
14.4 Which sex should compete most intensely for mates? 243
14.5 How male and female settlement patterns influence mating systems 245
14.6 Summary 248
14.7 Epilogue 249

References 250

Author Index 264

Subject Index 268

1

Why dunnocks?

1.1 Beginning with the Reverend Morris

'Unobtrusive, quiet and retiring, without being shy, humble and homely in its deportment and habits, sober and unpretending in its dress, while still neat and graceful, the dunnock exhibits a pattern which many of a higher grade might imitate, with advantage to themselves and benefit to others through an improved example'. With these carefully chosen words, the Reverend F.O. Morris (1856) encouraged his parishioners to emulate the humble life of the dunnock, or hedge sparrow *Prunella modularis*.

A gentle stroll around an English garden in spring might well lead the casual observer to endorse this view. The dunnock is indeed unobtrusive, as it shuffles about under the bushes searching for tiny insects and seeds, and its plumage does not, at first, catch the eye, some identification guides concluding with apparent exasperation that it 'lacks any conspicuous features'. But a closer look will reveal that the sober and unpretending brown dress is subtly and beautifully streaked with darker markings on the back and flanks, with the head and neck a delicate shade of grey. A bird watcher who pauses for a while will hear the male's warbling song and a search in the hedgerow may reveal a moss-lined nest with a clutch of glossy, turquoise-blue eggs, or a brood of young with the adults hard at work nearby, collecting billfuls of prey. The impression gained might be one of harmonious cooperation, as the dunnocks raise their offspring among a succession of blooms, from the crocuses and daffodils which accompany the first young of the year to the final broods fledged among the sweet scents of roses (Fig. 1.1).

A change in focus from casual observation of the species to detailed recording of the lives of individuals reveals that this Garden of Eden view of dunnock life is very much mistaken. The Reverend Morris's recommendation turns out to be unfortunate: we now know that the dunnock belies its dull appearance, having bizarre sexual behaviour and an extraordinarily variable mating system. Had his congregation followed suit, there would have been chaos in the parish and the Reverend Morris's devotion to ornithology and the writing of his *A History*

(a)

(b)

(c)

Fig. 1.1 (a) A female dunnock incubating. Nests are built in dense bushes and hedgerows. (b) The glossy, turquoise-blue eggs are laid in a nest neatly lined with moss and hair. Photographs by W.B. Carr. (c) An adult feeding a brood of hungry chicks, about a week old. Photograph by Eric and David Hosking.

of British Birds would have been severely disrupted. With the dunnocks colour-ringed for individual identification, another visit to the garden reveals an unsuspected diversity within the species. In one territory there is a pair (presumably what the Reverend Morris had in mind for dunnocks). Next door, however, a female is mated to two males. In a neighbouring territory there is the opposite arrangement, a male paired with two females. In yet other territories two males share two or even three females.

Closer observation of behaviour reveals intense conflicts among individuals. During egg laying, one male guards a female closely and attempts to prevent other males from copulating with her. The female, however, tries to escape his close attention and encourages other males to mate. This provokes endless chases around the territory and sometimes a female might mate many times each hour with two males, each apparently battling for paternity of her brood. The act of copulation itself is extraordinary, with a male pecking the female's cloaca carefully for a minute or so before he mates. Parental care is also very variable; sometimes a female raises a brood alone, sometimes she has the help of one male, sometimes the help of two males.

These observations raise fascinating questions. Why does this one species have such a variable mating system? Why all the conflicts over matings? Why such a strange pre-mating display? Why does parental care vary? Are dunnocks peculiar or do other small birds have similar variability? The dunnock also has another interest to a biologist, namely it plays host to the cuckoo *Cuculus canorus*, and accepts a cuckoo egg in its nest even though it is strikingly different from its own eggs, and even though acceptance leads to the total destruction of the dunnock's own reproductive success because the young cuckoo, on hatching, ejects all the dunnock's eggs or young over the side of the nest.

Throughout the book I shall emphasise the general interest of these and other observations for theories of behavioural adaptation. I have tried to bear in mind the story of the child who was given a book on penguins for her birthday. When asked whether she enjoyed the book, she replied 'Yes, but it tells me more about penguins than I really want to know!' My aim, therefore, is to present sufficient detail on dunnocks to interest an ornithologist, but not so much that it submerges the issues of more general interest to a student of behaviour and evolution. In this opening chapter I shall explain why I have spent ten years studying dunnocks and why much of the book is devoted to a description of individual differences in behaviour and reproductive success. In Chapter 2, I shall then introduce the dunnock and my study area in more detail.

1.2 Species versus problem-oriented studies

Studies of particular species are often best for answering certain kinds of questions. For example, the great tit *Parus major* and collared flycatcher *Ficedula albicollis*, have proved to be inspired choices for studies of the evolution of reproductive rates. Most of the population will breed in nest boxes, allowing easy measurement of reproductive success, their high density and synchronous onset of breeding permits large-scale experimental manipulations of brood size, and their sedentary habits enable measurement of both an individual's lifetime reproductive success and that of its offspring (Kluyver 1951; Perrins 1965; Gustafsson and Sutherland 1988; J.M. Tinbergen and Daan 1990). On the Galapagos Islands, Darwin's finches show extraordinary variability in bill size and shape, and live in an environment with great fluctuation in food availability. They have provided a wonderful source for studying natural selection in action and the evolution of community structure (P.R. Grant 1986; B.R. Grant and P.R. Grant 1989). In similar ways, studies of gulls have been seminal for ideas on the causation and function of displays (N. Tinbergen 1953), and of scrub jays *Aphelocoma coerulescens* and acorn woodpeckers *Melanerpes formicivorus*, for ideas on how ecological conditions set the stage for cooperative breeding in birds (Woolfenden and Fitzpatrick 1984; Koenig and Mumme 1987). David Lack clearly regarded choice of species as equally important to choice of problem. When John Krebs arrived as a new research student at the Edward Grey Institute, and announced that he wanted to study the problem of what happened to non-breeders in a population, Lack asked him what species he would work on. Krebs replied that he had not yet decided on the species. Later, Lack introduced him to the rest of the group as 'the new student who has no idea what he wants to work on yet'! (Krebs 1984).

My interest in the dunnock, a little brown bird, and at first sight apparently rather uninspiring, stems from two sources. The first was the publication in 1977 of two important papers on parental care and mating systems. In one Maynard Smith (1977) presented a novel theoretical analysis of how different patterns of parental care and mating systems may emerge depending on whether it was stable for one sex, both sexes, or neither, to desert a brood. It had been realised previously, particularly in a pioneering paper by Trivers (1972), that sexual reproduction was not always a harmonious enterprise in which males and females cooperated to maximise their joint reproductive success, but rather that each individual should behave to maximise its own success, even if this is at the expense of its mate. However, Maynard Smith provided the first logical framework for thinking about the costs and benefits for each sex of desertion versus caring. One of his main points was that desertion should vary depending on the benefits of care by one versus two parents, and also on the opportunities for finding another mate. The second paper was by Emlen and Oring (1977) and it provided a framework for thinking about how ecological conditions might influence these various costs and benefits. Ecological conditions set the stage on which individuals play out their behavioural strategies. Certain conditions of food abundance and predator pressure, for example, may permit easier monopolisation of several mates and hence favour the evolution of polygyny.

Soon after, two other papers then appeared concerning the curious sex life of the dunnock (M.E. Birkhead 1981; B.K. Snow and D.W. Snow 1982). Bird watchers had long suspected that the dunnock had an unusual mating system (e.g., Campbell 1952) and these first two studies of colour-marked populations showed that mating combinations varied considerably, even within a small population, including pairs (monogamy), a male with two females (polygyny), a female with two or even three males (polyandry), and more complex combinations involving two or three males sharing several females (polygynandry). Another study by Warui Karanja (1982) revealed this same extraordinary variability. These three studies stimulated me to begin a long-term study of a larger population. Obviously, a species with such variability provided an ideal opportunity for investigating the conditions under which different mating systems emerge. The theoretical ideas in the papers by Maynard Smith and Emlen and Oring suggested that a detailed study of the costs and benefits of parental care for males and females, together with a consideration of the potential to gain extra mates, might help to explain the variability. Indeed, this approach had already proved fruitful in studies of fish breeding systems (Baylis 1981) and work on mammals (Downhower and Armitage 1971) and birds (Alatalo *et al.* 1981) had also indicated that it was important to recognise conflicts of interest to understand male and female parental investment and mating systems.

The current trend for research in behavioural ecology is first to think of an interesting idea and then of a species that will be good for testing it. This was certainly how I came to study dunnocks, though in my case Maynard Smith and Emlen and Oring deserve the credit for the interesting idea and M.E. Birkhead, the Snows and Karanja for discovering that the dunnock had a variable mating system! Detailed studies of particular species for their own sake have become less fashionable and are sometimes disparagingly labelled 'the aardvark approach'; if no one has yet studied aardvarks then this alone is an excuse for doing research on them. I think this view is mistaken. We only have to think of David Lack's robins or Margaret Morse Nice's song sparrows to remind us that some of the main ideas in behaviour and ecology have come from long-term studies done for the love of a species as much as for the love of the ideas. A detailed study of any species is likely to throw up some novel problem of wider interest. Once I had the initial excuse to begin the study of dunnocks, I then discovered more interesting new questions from the bird watching than from reading the theoretical literature.

1.3 Studying species-typical behaviour versus individual differences

Niko Tinbergen pioneered modern field studies of behavioural design. He showed that just as an anatomist can take a structure, such as a bone, and make measurements to elucidate its function, so an ethologist can study the function of a behaviour pattern. One of Tinbergen's main interests was to understand why different species behaved in different ways. He realised that to understand these differences you had to examine how behaviour was of advantage to the individuals of the species. So he studied individuals to reveal the significance of species-specific behaviour patterns, for example, removal of egg shells, fear of cliffs and cryptic behaviour (N. Tinbergen 1974). David Lack, too, was especially interested in species characteristics. His book on the robin *Erithacus rubecula* is all about what the species does, why it defends territories, what its lifespan is, and so on (Lack 1965). Likewise, Lack's interest in clutch size was to understand why selection has favoured a particular average clutch size in a species (Lack 1966), and his interest in breeding systems was to understand why some species are monogamous while others are polygamous (Lack 1968).

Although these remain interesting problems for study, much of modern day behavioural ecology is aimed at understanding individual differences within a population. Why, for example, do some individuals lay larger clutches, have larger territories or more mates than others? There are three main reasons for this shift in emphasis, away from studies of species-typical behaviour patterns towards individual differences within a population.

(a) Individual differences are the raw material for natural selection. Studying them can be useful both for studying the costs and benefits of traits and selection in progress (Grafen 1988). Thus, detailed studies of individual red deer have revealed the costs and benefits of fighting for harems in stags and the importance of birth season and birth weight for survival in calves (Clutton-Brock *et al.* 1982). Studies of individual Darwin's finches have revealed the way in which different beak sizes are adapted to different types of food and have also shown why changes in the population occur over time as the food supply changes (P.R. Grant 1986).

(b) There may not be a single 'best' design for a behaviour pattern in a species. Where the success of a strategy depends on what others are doing, the stable outcome of selection may be for there to be variability in the population with different individuals doing different things. For example, within a population some males may fight for mates, while others avoid contests and invest in search instead, with both strategies enjoying equal success (Hamilton 1979; Maynard Smith 1982).

(c) Inspired by the theoretical ideas of Trivers (1972, 1974), Maynard Smith (1977) and Parker (1979), field workers have increasingly appreciated conflicts of interest within populations, not only the long-recognised conflicts between rival males, but also those between male and female within a pair and even between parents and their own offspring. These ideas have had a profound effect on how we interpret behavioural adaptation. Twenty years ago naturalists were chided for supposing that traits evolved in relation to their advantage to the species as a whole, and learnt to think about costs and benefits to individuals within a population and how traits influenced individual success (Williams 1966*a*; Dawkins 1976). Now the focus has changed again from ideas about how 'well designed' individuals might be to what selection is expected to produce when there are conflicts of interest. Is one party likely to win—for example, parents exert control over offspring, or males over females—or will the outcome of conflicts of interest be a compromise? Charnov (1982) has called this change of approach 'selection thinking'.

1.4 Conflicts of interest

It is this last theme, conflicts of interest, which dominates this book. I have invested time in detailed quantification of individual behaviour and reproductive success not simply to test how 'well designed' an individual's behaviour is, in the sense of how well it maximises that individual's fitness, but also to try and understand the source of conflicts of interest in breeding groups. The three topics of general interest which emerge are as follows.

(a) *Sexual conflict and mating systems.* I shall show how the dunnock's variable mating system is generated as the outcome of a conflict of interest, with a male having greatest reproductive success with polygyny and a female having greatest success with polyandry. Such conflicts are likely to be widespread in animal mating systems. Why, then, do not more species exhibit the variability shown by the dunnock? Under what conditions will particular individuals gain their best option despite the conflicting preferences of others?

(b) *Parental effort and life histories.* Some dunnock broods are fed by a pair of adults (one male, one female), others by a trio (two males, one female). How do cooperating adults reach a stable 'agreement' over how much work each should do? This is a general problem, relevant to all cases where several individuals share a common task. I show by experiment how an individual's effort is sensitive to what others are prepared to do, and investigate the influence of paternity on male parental care.

(c) *The adaptiveness of behavioural design.* We can use DNA fingerprinting to measure maternity and paternity and so test how well an individual's behaviour promotes its own reproductive success. But the dunnocks themselves do not have such precise measures to guide their parental effort. Instead, their behaviour is often based on 'simple rules'. For example, I show that polyandrous males do not discriminate their own young in multiply-sired broods, but simply feed chicks in relation to their prior access to the female during the mating period. Experiments can reveal how good this rule is, in the sense of how well it maximises a male's reproductive success. In some cases the dunnocks' simple rules have maladaptive outcomes, for example they lead to the rearing of a cuckoo chick rather than a brood of their own young. How 'well designed' should we expect an individual's behaviour to be, given the use of simple rules and the widespread conflicts of interest within a species (males versus females, parents versus offspring) and between species (dunnocks versus cuckoos)?

1.5 Summary

Particular species are often best for answering certain kinds of questions. The dunnock's extraordinarily variable mating system, which includes monogamy, polygyny, polyandry and polygynandry, all occurring within small populations, provides an ideal opportunity for investigating the conditions under which different mating systems emerge.

Behavioural ecologists often focus on individual differences within a species, for three reasons.

1. They provide raw material for natural selection and studying them can help elucidate both the costs and benefits of traits and changes in the population over time.

2. There may not be one single best design for a behaviour pattern in a species.

3. They can reveal conflicts of interest within populations.

The three themes of the book all involve conflicts of interest, namely conflicts within dunnock populations concerning mating systems and parental effort and a conflict between dunnocks and cuckoos. These raise interesting general issues concerning the evolution of cooperation in societies and how 'well designed' we should expect individuals to be.

2

Study species and study area

2.1 The accentors

In Old English, 'dun' means dull brown and 'ock' signifies little. True to its name, the dunnock is the archetypal little brown bird. Although sparrow-sized and commonly called the hedge sparrow, it has a thin bill and is not a true sparrow, but rather one of 13 species of accentors of the family Prunellidae (Table 2.1), mostly montane birds which occur throughout the Palearctic. It would be most logical to change the common name of *Prunella modularis* to 'hedge accentor' but I shall use the more familiar 'dunnock', the name used in previous publications and in the most recent handbook (Cramp 1988).

2.2 The dunnock

A detailed plumage description can be found in Cramp (1988). The sexes are similar in appearance, though males are on average a little larger (Chapter 5) and, especially in the breeding season, greyer on the head, neck and breast. Females rarely sing, and when they do the song tends to be shorter and simpler than that of the male (D.W. Snow and B.K. Snow 1983). Two other features are also sometimes useful for sexing birds in the hand during the breeding season. Only females incubate and develop brood patches, and males have swollen cloacas, packed with sperm (see Fig. 6.15).

Range and habitat

The dunnock's distribution is shown in Fig. 2.1. Like most of the other 12 species of accentor, the dunnock inhabits dense scrub in Arctic and Alpine habitats, and this is presumably where it evolved (Cramp 1988). In Switzerland it reaches highest density in mountain forests of spruce *Picea* and larch *Larix* up to *c.* 2,200m, preferring scrub along the forest edge or in forest glades. It extends up to the scrub just above the tree-line, and in the Pyrenees I have seen it in dwarf rhododendrons and junipers at 2,000m, but it remains below the higher, more open rocky areas, favoured by Alpine accentors. In Russia it occurs up to the tree-line at 2,600m, again preferring the densest vegetation, including birch *Betula*, juniper and bramble *Rubus*. In Scotland and the English Lake District it breeds among junipers on mountains up to *c*. 500–700m (Barnes 1970).

Unlike any of the other accentors, however, the dunnock has also extended to lowland temperate habitats. In the north of its range (Scandinavia, Russia), it is mainly in spruce forest, especially open areas containing dense scrub or brush piles (Tuomenpuro 1989). However, it also inhabits dense scrub in mixed and broad-leaved woodland, especially in open glades, along woodland edges or along rivers and streams. In western Europe (Belgium, France, Germany, Britain) it has extended to lowland farmland, where it inhabits hedgerows and woodland edges, and is also particularly common in suburban parks and gardens (Bevington 1991). Such spread to the lowland areas and urbanisation has not occurred in all parts of Europe; for example, in Switzerland it is largely absent from rural hedgerows, parks and gardens (Praz 1980).

In Britain, the shift away from montane and coniferous habitats is most pronounced, having occurred at least 200 years ago, perhaps much more. Even in 1802, Montagu wrote that it 'seems to prefer situations near the habitation of man'. It is most common in a variety of lowland scrub, including farmland hedgerows, churchyards, railway cuttings, parks, gardens, new plantations, and also rocky cliffs and offshore islands with dense scrub (e.g., bracken and bramble). It is the second most common bird of farmland hedgerows and one of the eleven most ubiquitous birds of lowland woodland (Fuller 1982).

There is probably no other part of its range where dunnocks can be seen so frequently as in Britain, a point graphically illustrated by a story told long ago to J.D. Wood (*in litt.*) by the pioneer field-ornithologist H.G. Alexander. Alexander was once taken to the woods near Berlin by a German ornithologist to find a black woodpecker *Dryocopus martius*, a scarce species which had long eluded him. They were hot on the track of one calling just ahead of them, when a dunnock sang near by. The German guide was so excited by this rarity that he turned aside to look for it. Meanwhile, the black woodpecker disappeared and Alexander had to wait another twenty years before he saw one!

We can only speculate about the cause of the dunnock's dramatic spread into lowland and urban areas in Britain. Most of Britain was covered in primeval woodland 6,000–8,500 years ago. This was cleared in the neolithic period (4,000–6,500 years ago), when forests were converted to farmland, but even by the early Iron Age (2,500 years ago) about half of England was probably still covered in wildwood (Rackham 1986). It is possible that the dunnock did not become a really common lowland species until extensive forest clearance, within the last thousand years, created the hedgerows and scrubby habitat which it prefers. The spread then may have been similar to that now occurring in Finland where a marked population increase has coincided with new forestry practices. The dunnock avoids pure coniferous forests with no scrub layer. Recently, foresters have created a more favourable habitat for the species by opening up

Table 2.1. Family Prunellidae: the accentors.

General characteristics. 13 species in a single genus *Prunella*, widespread throughout Europe and Asia with the greatest number of species in the mountains of central Asia. Small oscine passerines (suborder Passeres), generally brown or grey in colour, often streaked above. Some species also have rufous coloration, some have black or black and white spotted or barred throat patches, some have pronounced pale or white eyestripe. Sexes similar. Eggs immaculate blue or blue-green. Mainly terrestrial, often feeding close to scrubs and boulders. Food mainly insects in summer, with some seeds. Seeds often taken in winter (Cramp 1988).

Taxonomic affinities. Much debated. Bill shape and musculature thrush-like, but tarsus scutellated and syringeal musculature different from thrushes. Presence of crop, muscular gizzard and operculum covering nostrils suggest affinities with finches (Fringillidae) and buntings (Emberizidae) but may reflect convergent evolution rather than ancestry. Analysis of egg-white proteins suggests *Prunella* closer to Sylviidae (warblers) and Muscicapidae (flycatchers) than thrushes (Turdidae) (Sibley 1970). However, comparison of single-copy DNA sequences of the dunnock and those of other oscines suggests that *Prunella* is not closely related to any of these three groups, but rather is most closely related to the weavers (Ploceidae), with the next nearest relatives being pipits and wagtails (Motacillidae), sparrows (Passeridae) and finches (Fringillidae); these being grouped together in a superfamily 'Fringilloidea' (Sibley and Ahlquist 1981).

Species. Mostly little-studied, so the following information on range and habitat is approximate only (from Vaurie 1959; C.J.O. Harrison 1982; Cramp 1988). Three groups can be recognised: (a) Two large (*c.* 30–40g) montane species (1 and 2 below) which occur in mountains well above the tree-line. (b) Eight smaller species (*c.* 15–20g) occupying scrub around the tree-limit on mountains, or at the edge of northern forests at high latitudes (3–10 below). Some of these look very alike, with dark patches on the throat, face, or both, and a pale supercilium. They may be derived from a recent ancestor after geographical isolation. (c) Three rather similar dull-coloured, and geographically isolated, species (*c.* 20g) which can be regarded as a 'dunnock superspecies' (11–13 below), inhabiting dense scrub, sometimes in forest, both on mountains and (13 only) also at lower altitudes.

1. Alpine accentor *Prunella collaris*. Mountains, from NW Africa and central and southern Europe, eastwards to the Caucasus, Iran, Himalayas, China and Japan. Breeds above the tree-line and up to the snow-line, from *c.* 1,800m to *c.* 2,600–3,000m in the Pyrenees, Alps and Caucasus, but up to 4,000m in central Asia and even higher in the Himalayas—seen at nearly 8,000m on Mount Everest. In winter, usually shifts to lower slopes (often remaining above 1,800m).

2. Himalayan accentor *Prunella himalayana*. Mountains above the tree-line from east of Lake Baikal and NW Mongolia to Chinese Turkestan, north east Afghanistan and eastwards through the Himalayas to Sikkim.

3. Robin accentor *Prunella rubeculoides*. Dwarf rhododendrons, willows and sedges on open, damp, mountain meadows in the Himalayas and on the Tibetan plateau, between *c*. 3,600 and 5,600m.

4. Rufous-breasted accentor *Prunella strophiata*. Scrub (junipers and rhododendrons) above tree level, down to upper levels of open forest; Afghanistan, western China, Himalayas from west Kashmir eastwards to north Burma, between *c*. 2,400 and 4,500m.

5. Brown accentor *Prunella fulvescens*. Scrub in rocky mountain region of north Afghanistan eastwards to Himalayas and mountains of central Asia, between 3,000 and 4,300m.

6. Siberian accentor *Prunella montanella*. An inhabitant of the most northern forests on earth—along the northern limits of the mixed and coniferous forests of Siberia, and in the mountains of Siberia. Breeds along river banks and valleys and, on mountains, to tree-line. Winters in China (Knystautas 1987).

7. Kozlov's accentor *Prunella koslowi*. Thin scrub in semi-desert and on dry slopes, sparsely covered with grass and bushes, in the mountains of Inner and Outer Mongolia.

8. Black-throated accentor *Prunella atrogularis*. Birch forests and dense bushes (juniper) in the sub-Alpine zone and upper limits of the coniferous forests in Russia and China. Migrates in winter to lower altitudes, and south to Iran and western Himalayas.

9. Radde's accentor *Prunella ocularis*. Mountains of central Turkey and east through Caucasus, Armenia and Iran. Breeds at *c*. 2,000–3,000m on rocky mountain slopes with dense scrub (junipers).

10. Yemen accentor *Prunella fagani*. Restricted to the high mountains of Yemen. Sometimes considered a race of *P. ocularis*.

11. Japanese accentor *Prunella rubida*. Dense cover among pines and birches in mountains of Japan, up to tree-line.

12. Maroon-backed accentor *Prunella immaculata*. Thick undergrowth in dense coniferous forest, often along streams, in Himalayas from Nepal eastwards, *c*. 2,900 to 4,600m.

13. Dunnock, or hedge accentor *Prunella modularis*. Europe eastwards to the Ural mountains and south to Caucasian region and northern Iran. Northern populations migrate south for winter to central and southern Europe. Birds in mountainous regions move to lower altitudes. Varied habitat including scrub up to and just beyond tree-line in mountains (up to 2,500m) and also lowland habitats including spruce forests (especially those with dense cover, such as junipers or brush piles), dense scrub in open mixed woodland, shrubs and hedgerows in farmland and suburban gardens.

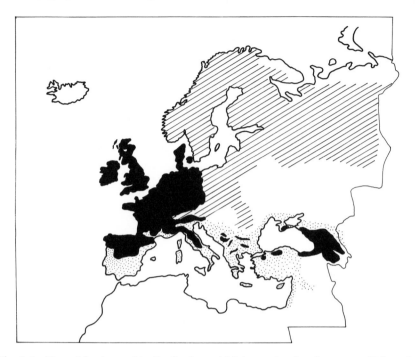

Fig. 2.1 Map of the dunnock's distribution, which is restricted to the western Palearctic (based on C.J.O. Harrison 1982 and Cramp 1988). Black areas = range where dunnocks are year-round residents (both breeding and wintering). Hatched areas = range where dunnocks are summer visitors only (breeding season). Stippled areas = range where dunnocks are winter visitors only (non-breeding season).

forest clearings and leaving dense brush piles, as well as by establishing new spruce plantations which also provide dense cover. In the late 1940s there were *c*. 10,000 pairs in Finland but this had increased to *c*. 250,000 by the mid 1970s and to 400,000 pairs by the late 1980s (Järvinen and Väisänen 1978; Tuomenpuro 1990).

A similar remarkable spread has occurred in New Zealand, where the dunnock was first introduced in 1868 and where it has successfully colonised all kinds of cover from sea-level to 1,600m, including saltmarsh, gardens, pine plantations and woodland scrub (Kikkawa 1966). It has also spread to most offshore islands and into the sub-Antarctic as far as Campbell island (Falla *et al.* 1970).

Foraging and diet

Typically, the dunnock is to be seen creeping about amongst dense vegetation in search of small insects and seeds. Most of its food is taken from the ground by gleaning or turning over small leaves. The diet consists largely of small invertebrates (e.g., beetles, snails, spiders, flies, earthworms and springtails) together with a variety of small seeds, taken especially in autumn and winter (Fig. 2.2; for a detailed list see Cramp 1988). The main feeding method is to hop along the ground, with body horizontal, making ceaseless pecking movements. It readily takes artificial food supplies, especially small bread crumbs, and long after other species have left a feeding area, the dunnock remains, picking up the tiny, almost invisible, scraps which other birds find unprofitable. It usually feeds unobtrusively under garden hedges or under a bird table and only occasionally ventures out on to raised surfaces.

2.3 The study area

My study area is the Cambridge University Botanic Garden (Fig. 2.3), where the breeding population of *c.* 80 dunnocks has been colour-ringed since 1981. It would be nice to think that Henslow walk, which runs through the centre of the Garden, was where Henslow and his famous pupil, Charles Darwin, used to go on their regular strolls; as an undergraduate in Cambridge, Darwin was often referred to as 'the man who walks with Henslow' (F. Darwin 1887). However, the Garden was not developed until well after Darwin left Cambridge.

J.S. Henslow was appointed to the University Professorship in Botany in 1825. One of his first tasks was to find a new site for the Botanic Garden. The old site established in the centre of Cambridge since 1762 was very restricted in size and being surrounded by old buildings caused a problem 'whereby the free circulation of air is impeded and obstructed'. Its proximity to old buildings gave rise to another, peculiar, hazard; numerous jackdaws *Corvus monedula* nested in the nearby church towers and chimneys of the colleges and they used to collect, as nest material, the sticks bearing the plant labels! One old chimney in Free School Lane contained eighteen dozen of them (Walters 1981). In 1831 a new site of 40 acres was obtained, less than a mile south of the city centre, but it was not opened formally until 1846 and half of the site has been planted only during the past 40 years.

Henslow's idea was to provide a range of living plants, including an extensive collection of trees, for service of the science of Botany. This contrasted with the older use of botanic collections which were often mainly of herbs, particularly those of medicinal importance. In the first decade of the 1900s, the Garden was used not only for studies of taxonomy, but also for experimental studies in the new and fast developing sciences of genetics and ecology. William Bateson cultivated plants for genetical studies and A.G. Tansley, one of the founders of ecology, and his students conducted experiments on competition between closely related plant species (Walters 1981).

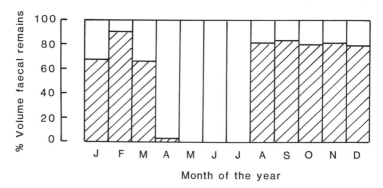

Fig. 2.2 Seasonal changes in the diet of dunnocks in a hedgerow habitat in Shropshire, England, studied from remains in faeces (*n* = 202 faecal samples) by Bishton (1986). The proportion of seeds (hatched) and invertebrates (white) in the diet was measured as the mean percent volume of the faecal sample which consisted of seed or invertebrate remains. Clearly, seeds were taken most in winter; mainly nettle *Urtica dioica*, Yorkshire fog *Holcus lanatus*, elder *Sambucus nigra*, dock *Rumex obtusifolius*, rosebay willowherb *Chamaenerion angustifolium*, meadow grass *Poa annua* and thistle *Cirsium vulgare*. Invertebrates were the main diet in summer, including beetles (Coleoptera; especially weevils Curculionidae, rove beetles Staphylinidae, ground beetles Carabidae and leaf beetles Chrysomelidae), spiders (Araneae), false scorpions (Pseudoscorpiones), snails (Gastropoda), earthworms (Oligochaeta), flies (Diptera) and springtails (Collembola).

Fig. 2.3 Part of the Cambridge University Botanic Garden, where there is a breeding population of *c*. 80 dunnocks. The birds feed mainly on the ground among the rockeries and nests are built in hedges and shrubs.

Unwittingly, Henslow's plan has also produced a wonderful site for studying the behaviour of garden and woodland birds. The Garden provides a mosaic of the various vegetation types favoured by dunnocks in more natural habitats, including areas of open woodland with dense undergrowth, hedgerows (mainly hawthorns *Crataegus* and various evergreens, e.g., *Taxus*, *Thuja*), areas of shrubs (e.g., Junipers) and rockeries with boulders and shrubs. The Garden is only 16 hectares in area yet contains about 80 breeding dunnocks, and dense populations of other species, including 100 pairs of blackbirds *Turdus merula*, 25 pairs of song thrush *Turdus philomelos*, 25 pairs of robins *Erithacus rubecula*, 30 pairs each of greenfinch *Carduelis chloris* and linnet *Acanthis cannabinna*, together with smaller numbers (1−10 pairs each) of breeding great tits *Parus major*, blue tits *Parus caeruleus*, coal tits *Parus ater*, long-tailed tits *Aegithalos caudatus*, tree-creeper *Certhia familiaris*, wren *Troglodytes troglodytes*, blackcap *Sylvia atricapilla*, lesser whitethroat *Sylvia curruca*, willow warbler *Phylloscopus trochilus*, goldcrest *Regulus regulus*, spotted flycatcher *Muscicapa striata*, starling *Sturnus vulgaris*, chaffinch *Fringilla coelebs*, bullfinch *Pyrrhula pyrrhula*, goldfinch *Carduelis carduelis*, redpoll *Acanthis flammea* and house sparrow *Passer domesticus*.

I have listed these simply to show what a rich bird life can exist so close to a city centre, provided a suitable habitat exists. In the Garden the dunnocks face the same hardships as in wilder places, including starvation and predation. Predators on the adults include kestrels *Falco tinnunculus* (a pair breed near the Garden and regularly hunt small birds there), tawny owls *Strix aluco* (1−3 pairs breed in or near the Garden), and occasionally hobbies *Falco subbuteo* and sparrowhawks *Accipiter nisus*. As well as these 'natural' predators, up to 10 domestic cats regularly patrol the Garden and pose a constant threat. Nest predators include carrion crows *Corvus corone* (1−2 pairs breed), jays *Garrulus glandarius* (occasional visitor) and grey squirrels *Sciurus carolinensis* (about 50 resident), as well as the cats.

In summary, although the study site is artificial, it provides the same habitat-type and selective pressures as exist in more natural places (see Section 3.8). In many ways the Botanic Garden is no more artificial than the other main habitats preferred by dunnocks in Britain, namely, farmland and woodland, both also extensively modified by man. The one major difference from wilder places is the greater breeding density and I shall discuss later whether this is likely to have influenced my results (see Chapter 3).

2.4 A colour-ringed population

This study would have been impossible without colour-ringing the population so that every bird was recognisable individually. Adults were caught by mist-net (Fig. 2.4) and given a unique combination of two colour rings on one leg, together with a numbered metal ring on the other (Fig. 2.5a). Nestlings were

(a)

(b)

Fig. 2.4 Mist nets were used to catch the adults. (a) Ben Hatchwell erecting a net, which is suspended between two bamboo poles. (b) The birds flew into the fine mesh and were caught harmlessly in pockets formed by shelves in the net.

ringed when they were halfgrown, 6–8 days old in the nest, and so the origins of all the young born in the Garden were known (Fig. 2.5b). Immigrants to the population, young birds born outside the study area who settled in the Garden to breed, were caught before the start of the breeding season, mainly from January to March, so that the whole population was colour-ringed before the start of the breeding season in late March and early April. In total, 427 adults bred in the Garden in the ten breeding seasons from 1981 to 1990, including 221 males and 206 females. Observations on their behaviour and reproductive success form the basis for the data in the chapters which follow.

The statement that 'the population was colour-ringed' conceals much of the effort put into the field work. Adults were most easily caught at dawn, when the birds were active and the nets less visible in the twilight, but wind and rain sometimes made catching difficult and some elusive birds required several dawn raids before we finally caught them. To ensure all the nests were found, the Garden had to be visited almost daily throughout the breeding season, from late March to July. However, all this hard work was repaid handsomely, not only by the scientific interest of studying birds as individuals but also by the simple pleasure gained from following the lives of old friends.

2.5 The advantages of field experiments

Once a whole population has been colour-ringed, it is very tempting to leave it undisturbed in order to follow individuals throughout their lives. With long-lived species this may be particularly important because the costs and benefits of an individual's actions are often delayed and there is no way of measuring these realistically unless individuals are followed throughout their lifetime. As an extreme example, in the long-tailed manakin *Chiroxiphia linearis*, a sub-ordinate male may spend several years dancing on a lek together with a dominant male. Their joint display increases the attraction of females, yet most if not all of the matings are performed by the dominant male. Why, then, does the subordinate male cooperate for the other male's benefit? McDonald (1989) has shown that subordinates may gain a long-term benefit from cooperation because they inherit control of the site when the dominant male dies and females may return to a place where they have mated previously. Thus a subordinate male may establish his own future reproductive success by first enhancing that of the other male.

Dunnocks, by contrast, breed in the first year after their birth, most of them live for just one or two years and, as I shall show later, short-term measures of the costs and benefits of behaviour patterns give a good measure of reproductive success. Although I have followed many individuals throughout their lives and have investigated long-term consequences of behaviour, I have also done field experiments which have disrupted natural events. The advantages of experiments are twofold.

(a)   (b)

Fig. 2.5 (a) Colour-ringed adult and (b) 7-day-old nestling.

(a) First, by controlling variation ourselves by experiment, we can eliminate the possibility that another variable, correlated with the feature under study, is the cause or effect of an event. For example, two cooperating males in polyandrous mating systems defend larger territories and have increased production of fledged young compared with single males. Is this because two males are indeed better than one at territory defence and offspring care? Or could the differences reflect other variables, such as single males having better quality territories (hence smaller area needed) or poorer quality mates (who are themselves responsible for the lower reproductive success)? Confounding variables, such as these, can often be teased apart by statistical analysis. However, it is often difficult to be sure that all the relevant variables have been measured. Wherever possible I have also performed simple experiments. Thus, in this example removal of one of the polyandrous males showed that the larger territory size in polyandry was not due to better territory defence by two males, but the increased production of young was due to their cooperation. Field experiments are not a panacea for solving all problems. They sometimes raise problems of their own, for example manipulation of one factor may disrupt features other than those under investigation (removal of a male may also cause changes in female behaviour). Throughout I have tried to perform adequate controls for the experiments.

(b) The second advantage of experiments is that they can be used to increase the range of natural variation to create circumstances that rarely or never occur. For example, competition for space can be studied by recording what happens when an individual is removed by a predator. However, to increase sample sizes I have often supplemented such 'natural removals' by experimental removals. In all cases, birds were kept in captivity for a short period and then released back on their territories. In other cases, experiments have allowed me to investigate dunnock responses to stimuli which do not occur in nature, for example, black or white cuckoo eggs, or whole clutches of cuckoo eggs. These experiments follow in the footsteps of those by Niko Tinbergen (1972) who taught us the power of simple manipulations and championed the use of the field as a natural laboratory for the ethologist.

2.6 Arrangement of the book

This book is not a comprehensive monograph on the dunnock. Instead, I hope it will be read in the same spirit as a detective story. The puzzle is the dunnock's extraordinary breeding behaviour and variable mating system. The job of the nature detective is to understand the alternative options facing individuals, to assess their reproductive payoffs, and then to discover whether different mating systems might emerge as a result of individuals competing to maximise their reproductive success. Much of the detective work therefore involves quantifying behaviour and reproductive success. I have summarised these results in tables and figures but to avoid clogging the text with statistical analysis, I have used numbered superscripts which refer to details presented at the end of each chapter. It should be possible to read the book without detailed reference to all this quantitative data, so those who are interested in the story and do not want to be bogged down with numbers should feel free to skip the tables and figures. However, ultimately the story stands or falls depending on the fit between individual behaviour and reproductive success and so I have presented the detailed data to show how the ideas were derived and tested.

I begin with a description of the variable mating system. Chapter 3 summarises the frequencies of the different systems during the ten years of the study, 1981–1990, and shows how the structure of the breeding population is related to winter and summer mortality, immigration and dispersal of young born in the Garden. In Chapter 4, I show in detail how the different mating systems are produced by various patterns of overlap between male and female territories. Natural and experimental removals reveal how males and females compete for habitat and mates, and a feeding experiment investigates the influence of food on territory size, which is an important determinant of the mating system. In Chapter 5, I discuss why particular individuals end up in different mating systems, particularly the influence of body size, age and familiarity with a territory. Chapter 6 then considers the behaviour of males and females in the various mating systems and describes conflicts of interest concerning copulations. The following three chapters analyse the source of these conflicts. Chapter 7 shows how DNA fingerprinting can be used to measure maternity and paternity and Chapters 8 and 9 assess male and female reproductive success in the various mating systems. The main conclusion from the first part of the book is that the variable mating system is the outcome of conflicts of interest between individuals within the population who are each behaving to maximise their own reproductive success, often at the expense of others.

The next part of the book considers parental effort and shows that conflicts of interest also lie at the heart of cooperative ventures, such as males and females joining to feed a brood of young, even though there may be no visible squabbles like those observed during competition for copulations. Experiments reveal how an individual's parental effort varies depending on what others are prepared to do and I discuss the problem of why individuals do not cheat, by doing less than their 'fair share' of work (Chapter 10). Chapter 11 considers how males with two females should best allocate their matings and parental care and shows that with choices involving two mating efforts, two parental efforts or mating versus parental effort, males choose the option which leads to greatest reproductive success. Chapter 12 investigates the influence of paternity on male parental effort and tests, by experiment, how well a male's chick feeding rules promote his own reproductive success.

In Chapter 13, I discuss the intriguing relationship between dunnocks and cuckoos. Dunnocks accept cuckoo eggs, even though they are very different from their own. Experiments with variously coloured model cuckoo eggs show a complete lack of discrimination, in sharp contrast to other cuckoo hosts. Is this evolutionary lag or a peculiar constraint which applies only to dunnocks? Given the fine-tuned adaptive nature of other dunnock behaviour patterns, why this apparent example of bad design? In the final chapter I discuss the relevance of the dunnock study for general issues concerning mating systems, cooperation and conflict, and the design of individual behaviour.

2.7 Summary

The dunnock is one of 13 species of accentors in the family Prunellidae, mainly montane birds which occur throughout the Palearctic with the greatest density of species in the Himalayas. As well as inhabiting montane regions, the dunnock, unlike other accentors, has also colonised lowland scrub, including farmland, parks and suburban gardens. The dunnock forages on the ground where it usually creeps about amongst the vegetation in search of small invertebrates and seeds.

My study area is the Cambridge University Botanic Garden, an area of 16 hectares with a breeding population of about 80 dunnocks. Although the site is artificial, it provides the same habitats and selective pressures as exist in more natural places.

The population was colour-ringed so that individuals could be recognised. A total of 221 males and 206 females bred on the study area from 1981 to 1990, and observations on their behaviour and reproductive success provide the data for the book. In addition to observations of natural variation, field experiments were used to examine responses to changes in particular variables and to circumstances that rarely or never occurred naturally.

3

Population structure and the variable mating system

3.1 Territory overlap of males and females

The idea of small birds living in exclusive territories as male-female pairs is so familiar that it was difficult at first to abandon this preconception, even though prepared by the earlier studies of dunnocks. The first part of each breeding season was spent simply walking round the Garden, plotting the positions of individuals on maps. An example from one year is shown in Fig. 3.1.

Females defended territories against other females and clearly occupied exclusive areas with very little, if any, overlap between neighbours (Fig. 3.1a). Although neighbouring females rarely met, when they did they displayed along their territory boundary, fluffing up their body feathers and hopping parallel to each other for a minute or so before parting.

Likewise, males defended territories against other males. Often one male defended an exclusive territory against his neighbours but sometimes the territories of two males overlapped (Fig. 3.1b). Usually the overlap involved two males jointly defending one territory (e.g., in the lower right of the map, GYL with SMR and BRL with pdbr). Sometimes, however, the patterns of overlap were more complicated (e.g., in the centre of the study area GBR's territory overlapped that of two neighbouring males WML and RRR). When two males shared a territory, there was a clear dominance order with one male (alpha) able to displace the other (beta) from feeding sites and from the vicinity of a female. However, the two males shared song posts and were often seen together displaying on a territory boundary against neighbours. Where there was a boundary between two territories each defended by two males (e.g., YSBR plus GNL, and WBR plus WMR, top left hand side of Fig. 3.1b), all four males were often seen displaying together on the boundary. Interactions between males were much more frequent than between females and involved singing duels (females rarely, if ever, sang), displays and chases.

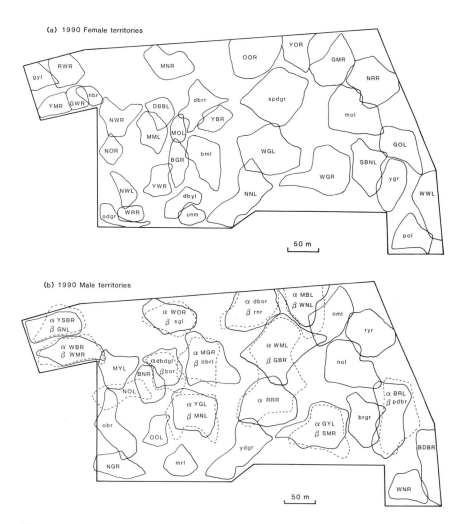

Fig. 3.1 Female (a) and male (b) territories at the start of the 1990 breeding season. In (b) the territories of males who defend territories alone, and of alpha males in shared territories, are indicated by solid lines and beta males in shared territories by dashed lines. Symbols refer to an individual's colour-ring combination, with lower case letters for birds born on the study area.

The variable mating system arose from the different patterns of overlap between male and female territories. Referring to Fig. 3.1, the following arrangements can be identified.

(a) *Monogamy*. One male territory overlaps one female territory, thus forming a pair. For example, in the top right hand corner of the map, male ryr and female NRR form a monogamous pair.

(b) *Polygyny*. A territory defended by one male overlaps two adjacent female territories. For example, in the lower left hand corner of the map, male NGR has two females WRR and odgr.

(c) *Polyandry*. A territory defended by two males overlaps one female territory. For example, in the lower right of the map males GYL (alpha) and SMR (beta) share female WGR.

(d) *Polygynandry*. A territory defended by two males overlaps two or more adjacent female territories. For example, in the lower right hand corner of the map, males BRL (alpha) and pdbr (beta) share two females GOL and ygr. In the top left hand corner, males WBR (alpha) and WMR (beta) share three females YMR, GWR and nbr.

These various mating systems are summarised in Fig. 3.2. Although the commonest forms of polygynandry were for two males to both share two or three adjacent females, there were also other ways in which territories over-lapped creating more complex patterns of mate sharing (Fig. 3.3). Throughout the first year of the study I found this variability bewildering. In retrospect, part of my bewilderment arose from giving each system a distinct label, following the strong predisposition of the human brain to put observations into distinct categories. I then began to worry about how to classify some of the less frequent cases of polygynandry illustrated in Fig. 3.3. For example, in (a) the system is certainly one of two males and two females (polygynandry) but one female is defended by just the alpha male while the other is defended by both males. Should the first female be classified as monogamous? It is different from normal cases of monogamy, because the male also associates with another female. Should the second female be classified as polyandrous? This again is different from normal cases of polyandry because the alpha male also associates with another female.

I remember cycling home one evening and suddenly realising what should have been obvious all along, that these various systems were simply examples of continuous variation on a single theme, namely an individual's ability to monopolise mates. I have arranged the mating systems in Table 3.1 to show this. Reading down the list, a male's access to females increases from no females to shared access to one female (polyandry), sole access to one female (monogamy), shared access to several females (polygynandry) and finally, best of all in terms of mating success, sole access to several females (polygyny). From a female's point of view, however, mating success increases from the bottom of the list to the top. Females have least access to males with polygyny, where they have to

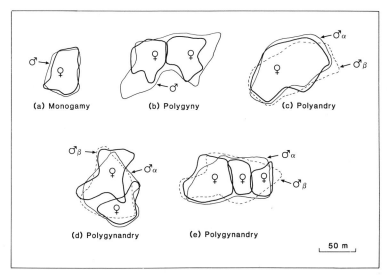

Fig. 3.2 How different patterns of overlap between male and female territories give rise to a variable mating system. Female territories are indicated by thick solid lines. A territory defended by one male (thin solid line) can overlap one female territory, creating monogamy (a), or two adjacent female territories, creating polygyny (b). A territory defended by two males (alpha = thin solid line, beta = dashed line) can overlap one female territory, creating polyandry (c), or two or three adjacent female territories, creating polygynandry (d and e).

share one male, and most with polyandry, where they have sole access to several males.

Observations showed that males were behaving so as to increase their access to females, while females were behaving so as to increase their access to males. There were thus frequent changes in the mating systems throughout the season as individuals moved to gain better options. This suggested that the variable mating system may be a reflection of different outcomes of sexual conflict. I shall present evidence for this view in Chapters 4 and 5. In this chapter I shall present data on the structure of the population, which contributes the players to the games which follow.

3.2 Frequencies of the different mating systems

Table 3.2 summarises the frequencies of the different mating systems at the start of the breeding season in the ten years of the study. In all years the most common systems were two males with one female (polyandry), one male with one female (monogamy) and two males with two females (polygynandry). There was

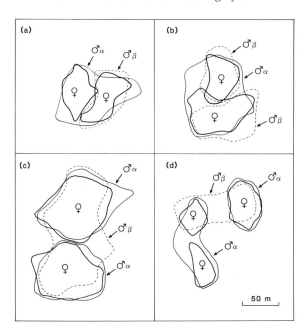

Fig. 3.3 Some unusual forms of polygynandry. Female territories indicated by thick solid lines, alpha males by thin solid lines and beta males by dashed lines. (a) Alpha male territory encompasses both females while beta male territory encompasses just one. (b) Alpha male territory encompasses both females but each female territory is also overlapped by a different beta male. (c) Adjacent female territories each overlapped by an alpha male, with a beta male encompassing both female territories. This is essentially two neighbouring 'pairs' with a beta male overlapping them both. (d) A neighbouring monogamous pair and polygynous trio, with a beta male extending from the pair to over-lap one of the polygynous females.

significant variation in the frequencies of the systems between years.[1] (Numbered superscripts refer to statistical analysis, presented at the end of each chapter.) On average, 37% of the males and 23% of the females were involved in polyandry, 27% of the males and 32% of the females in monogamy, 31% of the males and 38% of the females in polygynandry and only 3% of the males and 7% of the females in polygyny.

3.3 Population structure

The breeding population is divided into three categories in Table 3.3.

(a) *Surviving breeders from the previous year.* In England, the dunnock is remarkably sedentary. Although it ranks as the eleventh most frequently

Table 3.1. Arrangement of the various mating combinations to show how they reflect a continuum in the abilities of males and females to gain mates.

	Mating success	
Mating combination	For a male	For a female
Unpaired ♂	No females	—
Polyandry (e.g., 2♂ 1♀)	Share one female	Sole access to several males
Monogamy (1♂ 1♀)	Sole access to one female	Sole access to one male
Polygynandry (e.g., 2♂ 2♀)	Share several females	Share several males
Polygyny (e.g., 1♂ 2♀)	Sole access to several females	Share one male

ringed passerine bird in the scheme administered by the British Trust for Ornithology, with nearly half a million individuals ringed in the last fifty years (Mead and Clark 1989), it rarely features in the ringing reports and the vast majority of recoveries are within a few kilometres of where the bird was ringed (e.g., Spencer and Hudson 1977). A ringer after interesting recoveries would probably regard the dunnock as the species least worth ringing!

None of the birds ringed in this study were recovered more than 2km from the Garden. Once a bird settled on a breeding territory it remained very faithful to the site and the distance between the centre of its breeding territory from one year to the next was rarely more than 50m and usually much less.[2] Only three of the 427 ringed individuals in the study re-appeared after a breeding season's absence and all were on the edge of the study site and had probably moved only a short distance away. I think it is reasonable to assume, therefore, that when an adult disappeared it had died.

(b) *New breeders, born in the Garden.* All young born in the Garden were individually colour-ringed as nestlings. Of the 758 nestlings ringed during the nine years from 1981 to 1989, 75 stayed to breed. Of these, 74 were recorded breeding in the Garden during their first year. One turned up on the edge of the study area during its second year, probably having bred nearby in its first year. The young left their natal territories soon after reaching independence. Figure 3.4 shows that of the nestlings which

Table 3.2. Frequencies of the various mating systems at the start of each breeding season (early April). In 1981 only part of the Garden was studied, in subsequent years the whole Garden was studied.

Mating combination			Number of cases									
			1981	1982	1983	1984	1985	1986	1987	1988	1989	1990
Unpaired ♂			1	1	1	–	–	–	3	1	–	–
Polyandry	3♂	1♀	–	–	1	–	–	–	–	–	–	–
	2♂	1♀	5	14	12	6	7	6	8	7	6	3
Monogamy	1♂	1♀	12	10	8	12	6	7	7	10	21	15
Polygynandry	3♂	3♀	–	–	–	–	–	–	–	–	1	–
	3♂	2♀	–	1	–	–	1	–	1	1	–	1
	2♂	2♀	1	1	4	6	8	8	3	3	3	7
	2♂	3♀	–	1	2	2	1	–	1	2	–	1
Polygyny	1♂	2♀	2	–	–	2	1	–	1	1	2	2

Breeding population												
No. males			27	46	48	42	42	35	38	39	44	42
No. females			23	31	35	40	36	29	28	33	40	41
Total			(50)	77	83	82	78	64	66	72	84	83

remained to breed both males and females tended to move on average 3 to 4 territories away from their natal territory, a distance of some 250m. Given the small size of the study area it was not possible to gain a good measure of the distribution of dispersal distances because many of the young are likely to have bred outside the Garden.

(c) *Immigrants, born outside the Garden*. During the late summer there was a large influx of recently independent, unringed, juveniles originating from nests outside the study area. Unringed birds were not caught until the period just prior to the next breeding season (January–March) by which time they could not be aged for certain. However, given the sedentary nature of the adults and the fact that young born on the study area were ringed as nestlings, I think it is reasonable to assume that all these unringed birds were first year individuals, born outside the Garden the previous year. This is

Table 3.3. Structure of the breeding population at the start of each breeding season (early April). 1982 was the first year the whole population was ringed, so 1983 was the first year with known aged birds throughout the study area.

| Year | Total | | Surviving breeders from previous year | | Newcomers to breeding population | | | |
| | | | | | Immigrants (born outside Garden) | | Born in Garden previous year | |
	♂	♀	♂	♀	♂	♀	♂	♀
1981*	(27)	(23)	–	–	–	–	–	–
1982	46	31	(18)	(6)	–	–	–	–
1983	48	35	26	15	17	16	5	4
1984	42	40	21	18	14	18	7	4
1985	42	36	20	19	16	16	6	1
1986	35	29	24	13	8	14	3	2
1987	38	28	15	11	18	14	5	3
1988	39	33	16	11	18	18	5	4
1989	44	40	19	16	17	19	8	5
1990	42	41	25	21	10	15	7	5

*Only part of the Garden studied.

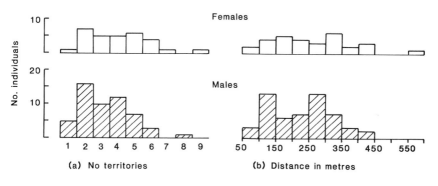

(a) No territories   (b) Distance in metres

Dispersal from natal to breeding territory

Fig. 3.4 Natal dispersal for 54 males (hatched columns) and 30 females (open columns). Data include some individuals which bred just off the study area. (a) Number of female territories from the nest where the bird was born to its breeding territory in the first year. 1 = next door territory, 2 = one territory between natal and breeding territory, etc. Mean for males = 3.26 (SE 0.21), for females = 4.00 (SE 0.34); t_{83} = 1.852, $P < 0.10$, 2-tailed. (b) Distance from nest where born to centre of breeding territory the next year. Mean for males = 235.9 (SE 12.4), for females = 263.7 (SE 21.2); t_{83} = 1.130, $P > 0.20$, 2-tailed.

Table 3.4. Survival of adults from start of one breeding season (early April) to the next and comparison of summer versus winter mortality. For sample sizes, see Table 3.3.

Year	% breeding adults who survived to start of next breeding season		% breeding adults who died:			
			During breeding season (April/July)		From August to start next breeding season (following April)	
	♂	♀	♂	♀	♂	♀
1981	66.7	26.1	18.5	34.7	14.8	39.1
1982	56.5	48.4	13.0	16.1	30.4	35.5
1983	43.7	51.4	16.7	17.1	39.6	31.4
1984	47.6	47.5	14.3	15.0	38.1	37.5
1985	57.1	36.1	7.1	22.2	35.7	41.7
1986	42.8	37.9	14.3	24.1	42.8	37.9
1987	42.1	39.3	13.2	21.4	44.7	39.3
1988	48.7	48.5	10.3	18.2	41.0	33.3
1989	56.8	52.5	15.9	17.5	27.3	30.0
1990	–	–	7.1	17.1	–	–
Mean	51.3	43.1	13.0	20.3	34.9	36.2
	$P = 0.05*$		$P < 0.01$		NS	

*Wilcoxon matched pairs tests, 2-tailed.

certainly true of the vast majority, which had pale pink tarsi (indicative of young birds—tarsi become darker with age), but it is possible that a very small proportion may have been adults who had bred close to the Garden the previous year.

Table 3.4 shows that about half the adults survived from the start of one breeding season to the next, with males surviving, on average, slightly better than females. There was no significant variation in either male or female survival in relation to the mating system (Fig. 3.5). The population was fairly stable during the study so about half of the breeders each year were first year birds, with a slightly (but significantly[3]) greater proportion among the females (mean for the 8 years in Table 3.3 = 56.3%) than among the males (mean = 49.5%). Most of the new breeders were immigrants to the population, with on average only 20–30% being born in the Garden the previous year. A greater proportion of the new male breeders were born in the Garden (mean for the 8 years in Table 3.3 = 28.4%) compared with new female breeders (mean = 17.3%), which were more likely to be immigrants.[4] This suggests that, as in

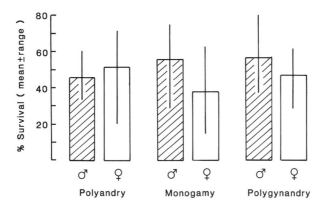

Fig. 3.5 Survival of adults from the start of one breeding season (1 April) to the start of the next. Histograms are mean and range for the nine years 1981-1982 to 1989-1990. There were no differences across mating systems in survival to the next year for either males (hatched columns) or females (open columns) (polygyny omitted because of the small sample size). Friedman 2-way ANOVA for males, $\chi^2_2 = 0.72$, $P > 0.30$; for females $\chi^2_2 = 1.156$, $P > 0.20$). Data for alpha and beta males were combined because there were no differences in their survival within either polyandry or polygynandry.

most passerines (Greenwood and Harvey 1982), young females dispersed further than young males, a result hinted at in Fig. 3.4.

3.4 Breeding season mortality

Of the adults present at the start of one breeding season who died before the start of the next, 29.3% of the males and 35.9% of the females died during the four months of breeding (April–July; figures are the mean percentage for the 9 years from 1981 to 1989 in Table 3.4). Thus adult mortality per month was as great during the breeding season as during the rest of the year. In the breeding season most of the mortality was probably caused by predators. The main culprits were cats. A pair of kestrels also frequented the Garden during the summer and hunted small birds in a manner more typical of sparrowhawks *Accipiter nisus*, by surprising their victims from low patrolling flights over the lawns or by pouncing on them from a perch. Tawny owls also certainly took some dunnocks; two rings were recovered in pellets.

During the breeding season, females were more likely to die than males (Table 3.4) probably for two main reasons. First, females did all the incubation and brooding of young and some were killed on the nest (probably by cats). Second, incubating females were easily recognised because they fed at a noticeably higher rate when they came off the nest and were probably less vigilant and so more susceptible to surprise attacks by predators.

3.5 Winter mortality

Mortality and winter severity

In winter, mortality was most likely caused by a combination of predation and starvation. When food is scarce, not only are small birds more likely to starve to death but they also take greater risks in foraging and so become more susceptible to predation (noticeable, for example, by their greater 'tameness' in cold weather). Figure 3.6 shows that female winter mortality increased with the severity of the weather but there was no such effect for males.

Dominance at feeding sites

Behavioural observations showed why females suffered increased mortality during harsh weather. In winter dunnocks feed mainly solitarily, with birds that bred the previous summer remaining largely on or close to their breeding territories. However, it was common for several individuals to aggregate temporarily at rich feeding sites, particularly in cold weather. Thus birds from the same polygynandrous or polyandrous breeding system may feed together, or two neighbouring pairs may associate at a food patch. Observations at these feeding sites, either natural sites or those created experimentally by provision of bread crumbs or softbill food, showed that there were linear dominance hierarchies, with males dominating females (Table 3.5), a result also found in M.E. Birkhead's (1981) study.

The effect of the hierarchy on subordinates was twofold. First, they appeared to take greater risks when foraging. At one feeding site frequented by eight individuals during a week of cold weather, I recorded who was the last to leave during a period of disturbance (by either a predator or myself) and who was the first to return afterwards. Table 3.6 shows that subordinates were much more often last to leave and first to return. I have no quantitative data to show how this influences predation but individuals who take greater risks are certainly likely to be most vulnerable. I once saw a kestrel attack a group of house sparrows *Passer domesticus* feeding in the Garden near the lake; as it approached they flew into the middle of a nearby tree but the kestrel caught one of the stragglers. On another occasion a blackbird *Turdus merula* was feeding out on the lawns, well before dawn, when a tawny owl swooped down from a pine tree above and killed it. The early bird may well be more likely to get the worm, but at the risk of increased predation, in this case from a nocturnal predator!

The second effect of the hierarchy was that subordinates were more likely to leave their territories during harsh weather and often left the Garden altogether to roam in search of food. During one cold spell from 8 to 28 December 1981 there was thick snow cover leaving just a few clear areas of earth and grass under the trees, where the dunnocks could search for food. On 16 of these days I monitored the presence of 29 males and 18 females who had been resident

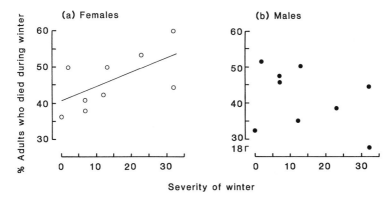

Fig. 3.6 Mortality of adults in relation to severity of the winter. Mortality refers to the percentage of breeding adults surviving to the end of the breeding season (1 August) who died during the winter (i.e., before the start of the next breeding season, 1 April). Severity of the winter was measured as the number of days (1 August–31 March) when there was snow or ice on the ground at 09.00 (from daily weather recordings in the Garden). (a) Female mortality increased with winter severity, Spearman rank correlation, $r_s = 0.687$, $P < 0.05$ ($y = 0.40x + 40.41$). (b) Male mortality was not related to winter severity, $r_s = -0.283$, NS.

Table 3.5. Examples of the linear dominance hierarchies at feeding sites in winter. The numbers in the table give the number of occasions displacements occurred, where one bird retreated when another approached.

(a) Two males and two females

Winners	Losers			
	$\alpha\sigma$	$\beta\sigma$	$\alpha\varphi$	$\beta\varphi$
$\alpha\sigma$	–	25	18	34
$\beta\sigma$		–	19	16
$\alpha\varphi$			–	10

(b) Three males and one female

Winners	Losers			
	$\alpha\sigma$	$\beta\sigma$	φ	$\gamma\sigma$
$\alpha\sigma$	–	20	27	31
$\beta\sigma$		–	23	11
φ				45

(c) Summary of dominance order at 11 winter feeding sites where there were males and females present.

No. birds	Dominance order	No. cases
2σ 2φ	$\alpha\sigma$, $\beta\sigma$, $\alpha\varphi$, $\beta\varphi$	4
3σ 1φ	$\alpha\sigma$, $\beta\sigma$, φ, $\gamma\sigma$	1
3σ 1φ	$\alpha\sigma$, $\beta\sigma$, $\gamma\sigma$, φ	2
2σ 1φ	$\alpha\sigma$, $\beta\sigma$, φ	1
1σ 1φ	σ, φ	3

Table 3.6. Influence of dominance hierarchy on who is last to leave during a disturbance and who is first to return to the feeding site afterwards. Data from a group of eight birds; the top four in the hierarchy were classified as 'dominants', the bottom four as 'subordinates'.

	No. occasions a dominant or subordinate:			
	Left ($n = 10$)		Arrived ($n = 13$)	
	First	Last	First	Last
Dominant	10	0	1	10
Subordinate	0	10	12	3
	$\chi^2_1 = 20.00$ $P < 0.001$		$\chi^2_1 = 12.76$ $P < 0.001$	

throughout the previous month. Each day, I provided food (bread crumbs and oats) on some of the territories so that 15 males and 7 females had access to extra food while 14 males and 11 females did not. Figure 3.7 shows two results. First, the presence of the extra food led to a marked increase in territorial attendance during the snowy spell. Second, in both feeder and nonfeeder territories subordinate birds were less likely to be present than dominants. Evidence that dominant birds caused exclusion of subordinates from restricted feeding sites was provided by two 'natural experiments' in the following winter. Two alpha males were killed by cats and in both cases the females subsequently spent more time at their feeders than before (one increased from 14% to 43% of its time at the feeder, the other from 0% to 21%). The exclusion of females by males from good feeding sites must also have contributed to their increased mortality in cold weather.

3.6 Sex ratio and variation in mating systems between years

As a result of increased mortality among the females in harsher winters there was a greater male-bias in the sex ratio of the breeding population following a hard winter (Fig. 3.8). Because so few males remained unpaired, the proportion of females involved in polyandry was closely related to the sex ratio (Fig. 3.9). Thus the severity of the winter and its effect on the sex ratio was partly responsible for the variation between years in the frequencies of the different mating systems (Table 3.2 above).

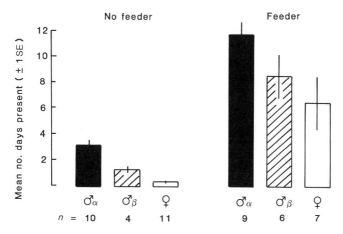

Fig. 3.7 The influence of the dominance hierarchy and the provision of extra food on territorial attendance during a period of 16 days continuous snow cover in December 1981. In the 'no feeder' category, alpha males (black columns) were present more often than beta males (hatched columns, $P < 0.01$) and females (open columns, $P < 0.001$), but the difference between beta males and females was not significant. In the 'feeder' category, alpha males were present significantly more often than females ($P < 0.05$) but other differences were not significant. Alpha males ($P < 0.001$), beta males ($P < 0.01$) and females ($P < 0.02$) were all more likely to be present if there was a feeder on their territory (*t*-tests, 2-tailed). Territories where there was only one male have been included in the alpha male columns.

However, even excluding polyandry, there was still significant variation between years in the relative frequencies of the other two most common systems, namely monogamy and polygynandry.[5] I am not able to explain this variation. It might be thought that polygynandry would be more common when population density was higher, or when female territory size was smaller, because dominant males could then more easily monopolise more than one female territory. However, there was no evidence for this.[6] Nor was the frequency of polygynandry related to the sex ratio.[6] It might be supposed that the frequency of polygynandry could be related to the proportion of old males in the population, older males being better able to monopolise mates (see later), but there was no such effect.[7] Finally, there was also no correlation with the degree of breeding asynchrony in females, which could influence the chances of a male being able to monopolise mates in succession[8] (see later).

3.7 Origins of individuals in the same mating system

In some species of birds, closely-related individuals share mates. For example, polyandry in Tasmanian native hens *Tribonyx mortierii* arises sometimes

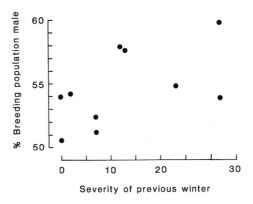

Fig. 3.8 The sex ratio of the breeding population ($n = 10$ years) was more male-biassed following a harsh winter; Spearman rank correlation, $r_s = 0.567$, $P < 0.05$. Winter severity scored as in Fig. 3.6.

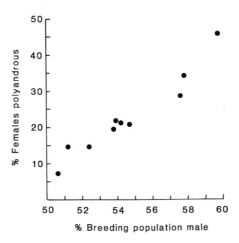

Fig. 3.9 The frequency of polyandry ($n = 10$ breeding seasons) increased with the proportion of males in the breeding population (Spearman rank correlation, $r_s = 0.948$, $P < 0.01$).

when two brothers share a female (Maynard Smith and Ridpath 1972). Polygynandrous breeding groups in acorn woodpeckers *Melanerpes formicivorus* are sometimes formed by a band of brothers, or father and sons, sharing a band of sisters, or mother and daughters, from another family (Koenig and Mumme 1987). The degree of relatedness between individuals is often the key to understanding the evolution of cooperation in animal societies (Hamilton 1964; Brown

Dominance hierarchy at a feeding site in winter. The alpha male (wing flicking) keeps the beta male (right) away from the food patch, a clearing in the snow. The female, third in line, awaits her turn on a branch nearby. Females are more likely to die in cold winters, so creating a greater male-bias in the sex ratio and increased polyandry the following spring.

1987), so it was important to discover the origins of the dunnocks in mating systems involving mate sharing.

Alpha and beta males in polyandry and polygynandry

There was no evidence that alpha and beta males who shared a territory containing one female (polyandry) or several females (polygynandry) were ever close relatives.

(a) In 6 cases both males were born on the study area and in all 6 they had different parents and were also not otherwise closely related. The distances between their nests were 90, 100, 130, 170, 260 and 290m.

(b) In 5 cases the older bird was born on the study area and the other was a first-year immigrant, born outside the Garden, and not a close relative.

(c) In 20 cases the older male was originally an immigrant while the other was a young bird born on the study area. In most cases the young bird came from a distant natal territory and in none was it a close relative of the older bird.

(d) In 6 cases both males were first-year breeders, but one was born in the Garden the previous year, while the other was an immigrant, born outside the Garden.

Thus in all 37 cases where the origins of both males were known, there were no instances of close relatedness. In the other cases it is also extremely unlikely that alpha and beta males were relatives. For example, in 26 cases the males were both immigrants but one was an old bird and the other a first-year breeder. They would only be likely to be relatives if birds born in one area independently chose to disperse to a particular area to breed, so that (for example) young born to the same parents in successive years ended up on the same breeding territories. There was no evidence for such a dispersal pattern; young born on the study area dispersed in all directions from a given site. This also makes it unlikely that relatives were involved in cases where both males were immigrants and of the same age ($n = 25$).

Females in polygyny and polygynandry

Likewise there was no evidence that the two females on adjacent territories who shared one male (polygyny) or several males (polygynandry) were ever closely related.

(a) In 2 cases both females were born on the study area. The distances separating their nests were 320 and 350m.
(b) In 6 cases one was an old bird, born in the Garden, and the other was a first-year immigrant.
(c) In 10 cases one was an old bird, originally an immigrant, while the other was a young bird born on the study area. In all cases the young bird came from a distant natal territory.
(d) In 4 cases, one was a first-year bird born in the Garden the previous year, while the other was an immigrant first-year bird, born outside the Garden.

In none of these 22 cases were the females close relatives. In the other cases, where both females were immigrants ($n = 34$), they were also very unlikely to be relatives for the reasons discussed above.

In conclusion, it is almost certain that in none of the 88 cases were alpha and beta males close relatives and likewise in none of the 56 cases were mate-sharing females closely related. In short-lived birds like dunnocks, where the brood sizes are small and where the young disperse away from their natal territories, this is not a surprising result. Dispersal away from the natal area ensured that the young did not breed with their parents, who remained faithful to their breeding territories in successive years. For the small proportion of young which did breed in the vicinity of their natal territories, high breeding mortality ($c.$ 50%) made it unlikely that they encountered their parents anyway. Furthermore, the small brood sizes and dispersal in all directions made it unlikely that siblings joined in the same mating systems. The absence of relatedness between alpha and beta males will also become clear in the next chapter, where I show that unpaired or widowed males will opportunistically join other males, as beta males, if the chance arises.

3.8 Comparison with other dunnock studies

The big worry about working in an artificial habitat, created by man, is that the birds may be exhibiting behaviour selected under different ecological circumstances. If natural selection has seldom 'seen' the reproductive outcomes experienced under Botanic Garden conditions, then attempts to understand dunnock behaviour by measuring reproductive success on my particular study site may be doomed to failure.

There are three reasons for concluding that this is probably not a problem. First, the Botanic Garden offers the same kinds of habitat, food and predators that dunnocks encounter in other places (Chapter 2). Second, most dunnocks in Britain live in habitats extensively modified by man, such as woodland edges and farmland, and probably have done so for several thousand years (Chapter 2), so the Botanic Garden is not unusual in this respect. Third, other studies done in a variety of habitats have revealed the same extraordinary variability in dunnock mating systems, so the behaviour patterns observed on my study site are not unusual for the species.

Table 3.7 compares the results from my study with those of four other studies, two in large gardens, one in semi-natural woodland and scrub and one in birch and spruce forest. In all five studies there was a similar male bias in the sex ratio of the breeding population and also a variable mating system. In the four studies in Britain, breeding density and territory size was similar and in all there was monogamy, polyandry, polygynandry and occasional polygyny occurring within the population. In the forest site in Finland density was much lower, territory size was much larger and monogamy was relatively more common, though polyandry also occurred regularly and there was one instance of polygynandry. This suggests that there may be more diversity in dunnock mating systems at higher density. Dunnocks have increased markedly in Finland in recent years (Chapter 2) and even during Jari Tuomenpuro's study (pers. comm.) there has been a tendency for polyandry and polygynandry to increase in frequency as density increases and male territory boundaries come into more frequent contact. At higher densities it may be easier for dominant males to monopolise a second female and higher intruder pressure from unpaired males may be more likely to lead to polyandry (see Chapter 4).

3.9 Summary

Females occupied exclusive territories which they defended against other females. Likewise, males defended territories against other males. Sometimes a territory was defended by one male alone but sometimes the territories of two

Table 3.7. Comparison of five studies of dunnock mating systems.

Studies	M.E. Birkhead (1981)	Karanja (1982)	Present study (Ch.3&4)	D.W. Snow and B.K. Snow (1982)	Tuomenpuro (1989; pers. comm.)
Habitat	School garden	Botanic garden	Botanic garden	Semi-natural woodland and scrub	Birch and spruce forest
Location	Edinburgh, Scotland	Oxford, England	Cambridge, England	Buckinghamshire, Southern England	Heinola, Southern Finland
Years of study	1978–79	1979–81	1981–90	1976–81	1984–88
Area of study site (hectares)*	11	3.6	16	9	408
Density (males per hectare)	1	2.2	2.6	1.3	0.1
Male territory size (mean, hectares)					
One male defends	0.2	0.7	0.29	0.1–0.6	1.54
Two males defend	0.56	0.9	0.56		2.17
Sex ratio: % males in breeding population†	58.3	57.1	54.5	56.7	56.8

Resident/migrant	Resident	Resident	Resident	Resident	Migrant (summer visitor)
Mating systems (No. cases)**					
Unpaired ♂	3	2	7	–	24
Polyandry 3♂ 1♀	–	–	1	–	1
2♂ 1♀	4	4	74	11	13
Monogamy 1♂ 1♀	7	4	108	7	112
Polygynandry 3♂ 3♀	–	–	1	–	–
3♂ 2♀	–	5	5	–	1
2♂ 2♀	1	–	44	3	–
2♂ 3♀	–	–	10	1	–
Polygyny 1♂ 2♀	1	–	11	1	–

*1 hectare = 10^4 m² †Start of breeding season; total for all years combined. **Total for all years of number of cases at the start of the breeding season.

males overlapped and they both defended the area against rivals. Where two males shared a territory, one (alpha) was dominant to the other (beta).

A variable mating system arose from the different ways in which male and female territories overlapped: a territory defended by one male could overlap one female territory (monogamy) or two adjacent female territories (polygyny); a territory defended by two males could overlap one female territory (polyandry) or several adjacent female territories (polygynandry). Intermediate cases also occurred. In all years the most common systems were polyandry, monogamy and polygynandry.

About 50% of the adults survived from one breeding season to the next, with males surviving slightly better than females. Survival did not vary with the mating system. Females suffered greater mortality than males during the breeding season and were more likely to die during harsh winters, probably as a consequence of their low positions in dominance hierarchies at feeding sites, where they took more risks in foraging and were forced to leave during cold weather.

As a result, the breeding sex ratio was more male-biased after a harsh winter and the frequency of polyandry increased. The variation between years in the relative frequencies of monogamy and polygynandry remained unexplained.

There were no cases of close relatives involved in any of the mating systems; both sexes dispersed away from their natal territories.

Four other studies have also found a variable mating system within a dunnock population, so the variability on my study site is not unusual for the species. The frequency of polyandry and polygynandry may increase as population density increases.

STATISTICAL ANALYSIS

(Numbers refer to superscripts in the text)

1. Table 3.2. Considering the three most frequent systems (all cases of polyandry, monogamy and polygynandry), there was significant variation among years, $\chi^2_{18} = 35.76$, $P < 0.01$. Excluding 1982 and 1983, which were affected by a feeder experiment (Chapter 4), $\chi^2_{14} = 22.37$, $P < 0.10$.

2. For 96 males and 69 females which bred in the Garden in successive years, there was little difference in the mean distance moved by males (28.7m, SE 2.1) and females (24.1m, SE 1.8); $t_{164} = 1.669$, $P < 0.10$, 2-tailed.

3. Wilcoxon matched pairs test, 2-tailed, $P = 0.05$.

4. Wilcoxon matched pairs test, 2-tailed, $P = 0.01$.

5. Table 3.2. Considering the frequencies of monogamy and polygynandry in the ten years, there was significant variation, $\chi^2_9 = 17.26$, $P < 0.05$. Excluding 1982 and 1983, which were affected by a feeder experiment (Chapter 4) $\chi^2_7 = 15.96$, $P < 0.05$.

6. There was no significant correlation across the ten years between the proportion of females involved in polygynandry, as opposed to monogamy, and total breeding population size (Spearman rank correlation, $r_s = -0.429$), or female population size ($r_s = -0.329$), or male population size ($r_s = -0.350$), or the percentage of males in the breeding population ($r_s = 0.188$).

7. There was no correlation across the ten years between the proportion of females involved in polygynandry, as opposed to monogamy, and the percentage of the breeding population which were old males (second breeding season or more; $r_s = 0.238$).

8. There was no correlation across the ten years between the proportion of females involved in polygynandry, as opposed to monogamy, and the standard deviation of first egg dates (a measure of the spread of the onset of breeding in the population) $r_s = 0.167$.

4

Territorial behaviour: competition for habitat and mates

4.1 Influence of male and female territory size on mating system

In Chapter 3 I showed how the variable mating system arose as a consequence of the different ways in which male and female territories overlapped. I now turn to the question of what causes these various patterns of overlap. Why do different individuals end up in different mating systems?

The various patterns arise largely as a result of differences in male and female territory size. Territory size was calculated by drawing polygons around the map registrations for each individual. The estimate of territory size increased with the number of registrations and from 40 to 100 were needed to give a good measure. During the breeding season individuals only rarely left their territories, for example to collect nest lining or to bathe in a pond, and these obvious distant excursions were excluded when drawing the polygons.

Within a given mating system there was no significant variation among years in either male or female territory size,[1] so the data have been combined for all the years of the study. In polyandry, there was no difference between the territory sizes of alpha and beta males.[2] However, in polygynandry beta male territories were, on average, smaller than those of the alpha males.[3] This arose simply because beta males sometimes associated with only some of the females in polygynandry (see, for example, Fig. 3.3). For cases where two or more males shared a territory, the following analysis refers only to alpha male territory size. Table 4.1 summarises mean male and female territory size in the different mating systems. In both sexes territory size varied enormously. In males the smallest was only 861m^2, while the largest was 13,173m^2, a difference of over 15 fold. In females the smallest territory was 684m^2, the largest 13,604m^2, a difference of nearly 20 fold!

The main conclusion is that differences in both male and female territory size contribute to the variable mating system. Thus, as shown in Fig. 4.1,

Table 4.1. Territory size of males and females in the various mating systems. In polyandry and polygynandry the territory sizes refer to the alpha males.

Mating system		Territory size (m²)					
		Male			Female		
		Mean	(SE)	*n*	Mean	(SE)	*n*
Polyandry	3♂ 1♀	7,103	(1,278)	2	7,103	(1,278)	2
	2♂ 1♀	5,669	(412)	36	5,450	(416)	36
Polygynandry	3♂ 3♀	6,272	–	1	2,144	(969)	3
	2♂ 3♀	5,733	(1,082)	5	1,916	(312)	15
	2♂ 2♀	5,554	(561)	24	2,894	(319)	47
Polygyny	1♂ 2♀	5,470	(1,101)	9	2,322	(309)	18
Monogamy	1♂ 1♀	2,906	(173)	58	2,778	(166)	58

Comparing sizes with 2-tailed *t*-tests. *Males*: 2♂ Polyandry ($P < 0.001$), 2♂ polygynandry ($P < 0.001$) and polygyny ($P < 0.05$) all significantly larger than monogamy. No significant differences between 2♂ polyandry, 2♂ polygynandry and polygyny, nor between 2♂ polygynandry involving 2 versus 3 females. *Females*: Polyandry (2♂) territories significantly larger than monogamy ($P < 0.001$), polygynandry ($P < 0.001$) and polygyny ($P < 0.001$). No significant differences between polygyny, monogamy and polygynandry involving 2 females. However, within 2♂ polygynandry, female territories were smaller in 3 female systems than in 2 female systems ($P < 0.05$).

the difference between monogamy and polygyny arises because, on average, polygynous males defend a larger territory than monogamous males, with average female territory size being the same in the two systems. On the other hand, differences between polyandry and polygynandry arise largely as a result of differences in female territory size; the average territory size defended by two males is the same in polygynandry and polyandry, but female territory size is markedly larger in polyandry than polygynandry, and within polygynandry it is smaller in three-female systems than two-female systems.

Therefore, the differences in mating systems involving one male arise largely as a result of differences in male territory size—males with the greatest mating success are those with larger territories. By contrast, the differences in mating systems involving two males arise largely as a result of differences in female territory size—two males defend just one female in cases where female territory size is large and two, or even three females, where female territory size is smaller. However, there were particular cases, for example, where a male with a fairly small territory still enjoyed polygyny because his two females had unusually small territories. Likewise, there were cases where two males jointly

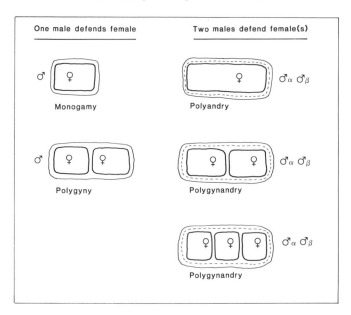

Fig. 4.1 Schematic diagrams of male and female territory sizes in the different mating systems, drawn to scale to indicate average sizes. Female territories indicated by thick lines, monogamous and alpha males by thin lines and beta males by dashed lines. These show how both male and female territory size influence the mating system. The difference between monogamy and polygyny is largely due to a difference in male territory size. The difference between polyandry and polygynandry is largely due to a difference in female territory size.

defended only a small territory but nevertheless encompassed two even smaller female territories.

The conclusion from this section is that we need to understand variation in both male and female territory size if we are to explain the variable mating system. Before examining the sources of this variation, I shall first discuss how males and females settle on their breeding territories in spring.

4.2 How the various mating combinations arise

The conventional view of pair formation in passerine birds is that the males first set up territories and advertise for mates by singing. Females are then supposed to choose between male territories and it is their choice that determines which males gain several mates (polygyny), which gain just one mate (monogamy) and which remain unmated (Orians 1969). This pattern seems most obvious in species that are summer visitors to temperate regions, where the males arrive on the breeding grounds a week or so before the females, and one of the joys of spring is to walk through the countryside serenaded by their songs. Females arrive later and may sample several male territories before settling to breed.

Male competition for female territories

In dunnocks, however, the formation of mating combinations does not follow this pattern. The movements of females are not constrained by male territory boundaries and females do not move about sampling male territories. In fact the reverse process occurs. Females appear to choose a breeding territory independently of male distribution and their settlement is largely determined by competition with other females for space. Males then impose themselves on the female distribution, and compete to monopolise the female territories. In the rest of this chapter I shall present observational and experimental evidence for this view. In the final chapter I shall suggest that many other species may follow the dunnock pattern and so the conventional view of females choosing among male territories may often be wrong.

Males and females which bred in the Garden the previous summer remained on or near their breeding territories throughout the winter, except during cold spells (Chapter 3). I do not have good data on the timing of settlement by first-year birds. Some certainly settled in the Garden during the autumn but others did not arrive until the period from February to early April. These newcomers in early spring seemed to be mainly females. During the winter, from two to six individuals shared overlapping ranges and there was little overt aggression except at rich feeding patches where they all sometimes congregated to forage

Two neighbouring males meet on the boundary between their territories and display, with their body feathers fluffed out. The female paired with the male on the right watches in the background. Males defend territories against males and females defend territories against females. Various mating systems arise depending on how the male territories overlap the female territories.

(Chapter 3). During mild spells in late January, and then in earnest in February and March, males began to sing and defend exclusive song territories. At first, a male tolerated the presence of other males feeding within his song territory but if they sang too, then they were chased. Later on, all trespassing males were chased whatever they were doing.

From late February onwards, females too began to remain more strictly within exclusive territories and a male was often seen chasing a female continuously for periods of up to ten minutes. During these chases the female would often fly around the whole of her territory pursued closely by the male, who followed her every move, twisting and turning through the vegetation. Whenever the female perched on a branch for a short period, the male sat by her and occasionally sang, but as soon as she flew off again he followed in hot pursuit. These sexual chases gave the impression that the male was learning the boundaries of the female territory and then attempting to set up an exclusion zone around her, advertised by song. Some males succeeded in defending one or more female territories alone, thus forming monogamy or polygyny.

Formation of polyandry

In other cases, however, two neighbouring males each set up a song territory in part of a female's territory (Fig. 4.2). Whenever this happened, each male attempted to follow the female when she crossed the boundary into the other male's territory. Chases occurred, and at first the dominance order of the two males would reverse as they crossed into the other's song territory, with each resident being dominant in his own half. However, after several days chasing and song duels, with both males pursuing the female over the whole of her area, a clear dominance order would emerge between the males and both then sang over the whole of the female's territory (Fig. 4.2). The result, therefore, was that the two neighbouring males coalesced their territories and came to share the female as alpha and beta male of a polyandrous trio. Once territory coalescence occurred, both males shared song posts with little aggression between them and they often defended the territory together by displaying as a team against neighbours.

The fact that females in polyandry have territories about twice the size of monogamous females (Table 4.1), is exactly what might be expected if polyandry was formed by coalescence of two single male territories, with two males being able to defend twice the area of one male. However, polyandry did not always form in this way. Some first-year males wandered around the Garden in early spring, singing briefly and being evicted by the resident male from each territory in turn. Some of these males left the study area, but some were successful in forcing themselves on to territories where there was a monogamous pair. The resident male always chased the newcomer and tried to evict it, but if it persisted the frequency of chasing declined and eventually it was tolerated as a beta male in polyandry. One such wandering male spent 19 days roaming

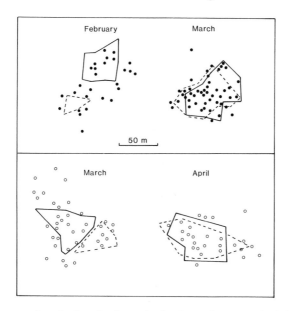

Fig. 4.2 Two examples of polyandry formation by the coalescence of neighbouring male territories. In each case two males have separate song territories in early spring, denoted by solid and dashed lines. The solid or open circles represent sightings of the female. By late spring the two males have coalesced their territories and have overlapping song polygons, sharing the whole of the female's range. From Davies and Lundberg 1984.

the Garden and was seen attempting to settle on nine different teritories before it was finally accepted on one, while another male spent 4 days wandering and was seen to be evicted from five territories before it managed to settle on another.

The formation of polyandrous trios at the start of the breeding season was followed in most detail during 1981 to 1983. Of the 25 cases studied, 19 formed by coalescence of neighbouring male territories, while 6 did so by wandering males forcing themselves onto pairs. Later on in the breeding season, when some of the females died, this second method of polyandry formation increased in frequency as the bereaved males joined neighbouring monogamous pairs (see below).

Formation of polygynandry

Polygynandry also formed in these same two ways (Fig. 4.3).

(a) *Coalescence of neighbouring male territories.* The most common cases of polygynandry were two males sharing two females. These often began as two neighbouring monogamous pairs, with each male restricting its territory to

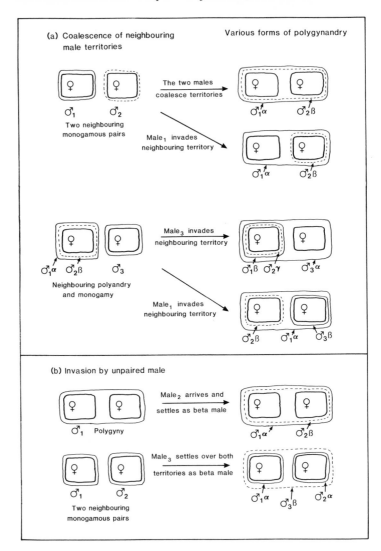

Fig. 4.3 The formation of polygynandry. See text for description of events.

encompass one female. One male then invaded the other female's territory, usually at the time she began her nest and was offering copulations, and he then took over the whole area as alpha male. The other male then became beta male and likewise usually wandered over both female territories so coalescence was complete (31 cases), but occasionally he remained in his original female's territory (3 cases; for an example see Fig. 3.3a).

Coalescence also sometimes occurred between neighbouring polyandrous and monogamous mating combinations (Fig. 4.3). In one case the monogamous male invaded the territory defended by the alpha and beta male, and took over their female. These two males then became beta and gamma male, but remained in their original territory. In another case the invasion was by the alpha male from the polyandrous system. The beta male remained with his original female and the original monogamous male remained as beta male with his female. The result, therefore, was an alpha male dominating both females, each also with a separate beta male (for an example see Fig. 3.3b).

(b) *Invasions by wandering males.* Polygynandry also occurred when an unpaired male forced himself onto polygyny, thus creating two males with two females. Sometimes a wandering male became beta male over two neighbouring monogamous pairs (as in Fig. 3.3c) or settled as beta male to a male with one female and also over part of the territory of an adjacent polygynous male (as in Fig. 3.3d).

4.3 Mating system changes due to mortality: natural experiments

The changes brought about during each breeding season due to mortality (predation) provided natural experiments which also gave a valuable insight into how the different mating systems formed. Table 4.2 shows that where the death of a female left the males without a mate (i.e., in monogamy and polyandry), the males usually immediately abandoned their territory and competed for other females on adjacent territories. By contrast, where the death of a male left a female without a mate, the female always remained on her territory and another male usually arrived to claim her. This important result confirms the observations in the previous section, showing clearly that males move to settle on the females' distribution, not vice versa. In seven cases one polygynous female died and the male stayed with the remaining female, forming monogamy. In 22 cases one female died in polygynandry, and the males again stayed with their remaining female, forming polygyny. Therefore the emigration by males shown in Table 4.2 was definitely a response to sudden loss of their only mate and was not simply a response to disturbance by a predator.

Movement by males was not only caused by loss of a mate on their own territory. Sometimes males moved in response to vacancies on adjacent or nearby territories. For example, of the nine cases where a monogamous male died (Table 4.2), in eight cases he was replaced by another male. (The one case of no replacement was in July, at the end of the breeding season.) Two of these replacements were unpaired males from next door, two were beta males from neighbouring polyandrous systems (who thus became monogamous, leaving their alpha male also as a monogamous male), two were neighbouring

Table 4.2. The results of natural removals, caused by predation during the breeding season, showed that males moved to monopolise female territories rather than females moving to settle in male territories.

Mortality	Remaining individual	Response of remaining individual	
		Moves to another territory	Stays on old territory
Monogamy (♂ ♀)			
Female dies	Male	11	1*
Male dies	Female	0	9†
Polyandry (2♂ ♀)			
Female dies	Alpha male	3	3*
	Beta male	3	3*

In monogamy, the difference between the responses of bereaved males and females is highly significant, $\chi^2_1 = 13.533$, $P < 0.001$.

*In these 4 cases the female died in late June–July, at the end of the breeding season. †In 8 cases another male arrived to defend the female—see text. In the remaining case the male died in July, at the end of the breeding season.

monogamous males who expanded to enjoy polygyny, one was a neighbouring alpha and beta male with one female who expanded to become polygynandrous, and one was a neighbouring alpha and beta male with three females who expanded to encompass a fourth mate. Table 4.3 summarises all the male movements in response to vacancies and shows that of the 33 cases, 32 led to an increase in male mating success with one movement resulting in no change. Clearly then, males were very quick to enlarge their territories or move to defend another territory altogether if this led to increased access to females.

Even with daily visits to the Garden during the breeding season, and even with all the adults colour-ringed, it was still a challenge to keep up with the constant changes in mating systems as males jostled for better opportunities. Some of the deaths caused interesting 'knock on' effects in neighbouring territories. Two examples are shown in Fig. 4.4.

Conclusion

The picture that emerges is one of females settling to form a mosaic of female territories (Fig. 3.1), with males then competing to monopolise parts of this mosaic by themselves setting up territories overlapping the female distribution. The variable mating system reflects different degrees of female defence by males, with all the patterns described in Chapter 3, both the common ones (Fig. 3.2) and more unusual ones (Fig. 3.3), forming part of the same continuum.

Table 4.3. When males moved their territory boundaries during the breeding season in response to vacancies caused by mortality, either by expanding their territories or by moving to a new territory, the result was an increase in their mating success. Mating status arranged in increasing order of male success. The number of cases of each change is shown.

Male mating status before moving	Male mating status after moving					
	Beta male in polyandry	Beta male in polygynandry	Alpha male in polyandry	Monogamous male	Alpha male in polygynandry	Polygynous male
No female	12	1	1	3	–	–
Beta male in polyandry	1	3	–	2	–	–
Beta male in polygynandry	–	–	–	2	–	–
Alpha male in polyandry	–	–	–	–	1	–
Monogamous male	–	–	–	–	–	2

Additional movements included: (a) a gamma male in 3♂ 2♀ to become a beta male in 2♂ 2♀; (b) an alpha male and a beta male in 3♂ 3♀ to become an alpha and beta male in 3♂ 4♀; (c) an unpaired male to become a gamma male in 3♂ 2♀; (d) an unpaired male to become a gamma male in 3♂ 3♀.

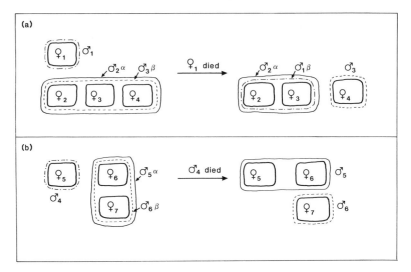

Fig. 4.4 Two examples of how the changes in mating systems caused by mortality can produce 'knock on' effects in neighbouring territories. (a) When female$_1$ died, her male (male$_1$) settled next door as a third male in polygynandry. With the increased pressure from this invader, the resident alpha male (male$_2$) remained with two of his females, female$_2$ and female$_3$, so the former beta male on this territory (male$_3$) became free to monopolise female$_4$ as a monogamous male. (b) A monogamous male (male$_4$) died and the neighbouring alpha male (male$_5$) expanded his territory to include the widowed female. His beta male (male$_6$) was left free to monopolise female$_7$ as a monogamous male, with the alpha male becoming a polygynous male with female$_5$ and female$_6$.

At the start of the chapter I showed that the variable mating systems were an outcome of differences in both female and male territory size. This now leaves us with two questions. First, what determines female territory size when females form their mosaic of exclusive territories? Second, what then determines male territory size when males, in turn, compete to monopolise the female territories? I shall consider these two questions in the following sections.

4.4 Factors influencing female territory size

Competition among females for space

Apart from areas of open lawn, which provided neither nest sites nor preferred feeding areas, the female territories formed a jig-saw pattern filling most of the Garden, with little space between them (Fig. 3.1). This suggests that females competed for space. This was examined by following the consequences of 34 vacancies caused by predation of females during the breeding season (Table

Table 4.4. Results of natural removals of females during the breeding season, caused by predation.

What happened to the vacancy?	No. cases female died				
	April	May	June	July	Total
New female arrived to fill it	1	–	–	–	1
Neighbouring females expanded to fill					
Whole area	4	–	–	–	4
Part of area	3	5	4	–	12
Vacancy remained unoccupied	3	3	6	5	17

The vacancy was more likely to be wholly or partly filled in the first half of the breeding season, April–May (13 out of 19 cases) than in the second half, June–July (4 out of 15 cases) $\chi^2_1 = 5.846$, $P < 0.02$.

4.4). Vacancies were more likely to be filled early in the season and in all but one case they were quickly taken over by neighbouring females.

This result shows that, at least once the breeding season has begun, there were few, if any, females excluded from the breeding population. If there were, they would have been expected to fill the vacancies or at least to compete for them. However, it does suggest that the females who settle restrict each other's movements. Of the 16 cases in Table 4.4 where neighbours took over at least part of the vacancy, 11 increased their territory size, some of them markedly, while 5 simply moved the position of their territories and had a net decrease in territory size. Thus female territories are sometimes restricted in either size or position by competition with other females. It was also noticeable that during incubation, when a female spent most of her time on the nest and so was less able to defend her territory, neighbouring females more often trespassed in search of food.

Paradoxically, the females who showed the largest increase in territory size as a result of expansion into a vacancy were those who initially had the larger territories.[4] Females with smaller territories tended to simply shift into the vacant area, with little change in their territory size or even a net decrease. A trivial explanation could be that small territories tended to occur together (see Fig. 3.1) and so expansion by females with small territories was more constrained. However, even considering the 12 cases in Table 4.4 where the vacancy was only partly filled (i.e., further expansion possible), there was still greater expansion by females with initially larger territories.[5] This suggests that it is the females with the largest territories who may be most constrained by neighbours. As I shall show below, the larger territories occurred in poorer habitats which may explain this result.

Influence of habitat

The Garden is laid out in patches of distinctive habitats ranging from areas of open lawn with hedges or scattered bushes to areas with denser vegetation, including patches of woodland with dense undergrowth (Fig. 4.5). It was very obvious that female territory size was smaller in areas with denser vegetation. In areas with the densest vegetation, average female territory size was less than half that in the most sparsely vegetated areas (Fig. 4.6). These differences between sites were consistent between years[6] and were certainly due to differences in habitat type, because within habitat categories there was little difference between sites. It is interesting that territories in the areas with dense bushes were not smaller when there was woodland canopy. This suggests that it is the ground vegetation which has the greatest influence on female territory size, which accords with the species' preference for low nest sites and ground foraging.

Are these marked differences influenced by food availability, or nest sites, or both? Nest site availability certainly had some influence on female distribution. For example, in one area of the Garden the only suitable nest site was a snow-berry bush (*Symphoricarpus rivularis*). This was dug up to get rid of a rabbit colony breeding in a warren underneath, and causing havoc with the vegetables in the nearby beds illustrating plant taxonomy! After the bush was removed, no female bred in this part of the Garden. In another area a hawthorn hedge (*Crataegus*) was cut back severely and no females bred there for two years, until the hedge had grown up again. However, these cases were unusual. In most of the Garden there seemed to be an abundance of nest sites and when hedges and bushes were pruned the birds simply nested in another site within their territory.

This suggested that the main cause of the habitat differences in territory size might be availability of feeding sites. In the Garden the dunnocks spent almost all their time feeding in dense vegetation, usually in flower beds, under bushes and hedges, or at the base of trees. These good foraging areas occurred in patches of various sizes interspersed among open areas of lawn which were rarely used for feeding. In some parts of the Garden the foraging patches were densely distributed whereas in other parts they were more widely scattered. Figure 4.7 shows that female territory size was largest in areas where there was a low density of suitable feeding patches and smallest where most of the territory consisted of dense vegetation and flower beds. Furthermore, the points in the graph roughly follow the expected curve if female territory size was adjusted so that each contained the same average feeding area.

Figure 4.8 shows, for these females, how variation in their territory size influenced their ease of monopolisation by males. It is clear that the larger a

(a)

(b)

Fig. 4.5 Habitat variation in the Garden. (a) Hawthorn *Crataegus* hedgerows and open lawns. (b) Rockery and scattered bushes (*Juniperus* in foreground) with denser bushes and woodland behind.

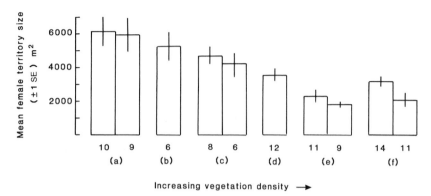

Fig. 4.6 Female territory size varied significantly among ten sites in the Garden with large patches of distinctive habitat. Numbers under the histograms indicate sample sizes. (Kruskal-Wallis 1-way ANOVA, $\chi^2_9 = 58.869$, $P < 0.001$.) Territory size decreased with increasing vegetation density (from left to right). (a) Open lawn with hedges or scattered bushes. (b) Open rockery with scattered bushes. (c) Open lawn with some dense bushes. (d) Long grass, hedges and scattered bushes. (e) Dense bushes. (f) Dense bushes with woodland canopy. There were no differences between sites within the same habitat categories, except between the two woodland sites (f), where the site with the smaller territories had denser ground vegetation ($P < 0.05$; Mann-Whitney U-test, 2-tailed).

female's territory, the more likely she is to be defended by more than one male. This suggests that a female's territory size is influenced by food distribution, and the size of her territory then determines the mating system through her ease of monopolisation by males.

4.5 Experimental changes caused by feeders

Predictions

In 1982 and 1983, Arne Lundberg and I tested this hypothesis by experiment (Davies and Lundberg 1984). In each year we put out food, every day from mid January to mid May, in the centre of ten randomly chosen female territories. The food consisted of a mixture of oats, bread crumbs, maggots, mealworms and a softbill food made from insects and seeds blended with honey and vegetable oil. We predicted that if food availability was the main determinant of female territory size, then females with feeders should have smaller territories than controls. Furthermore, if female territory size, in turn, influenced the mating system then feeder females should be more easily monopolised and so the mating system should change towards increased male success.

Birds with feeders quickly came to use the provisioned food and some changes in territory boundaries occurred so that feeders sometimes ended up on

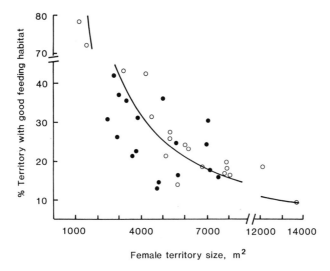

Fig. 4.7 Female territories were largest in poor feeding areas where there was a low density of flower beds and bushes. (●), data from 1981, Spearman rank correlation $= -0.526$, $n = 17$, $P < 0.05$. (○), data from 1982 (excluding feeder territories), Spearman rank correlation $= -0.873$, $n = 18$, $P < 0.01$. The curve drawn through the points is the expected curve if all females had the same feeding area within their territory, namely the average area of flower beds and bushes in female territories, which was 1269m^2. Note the breaks in both axes. From Davies and Lundberg 1984.

a boundary between two or three female territories. In 1982, twelve females had regular access to the ten feeders and in 1983, sixteen females did so. The food had a marked effect on time budgets. In February and March, before nest building began and when males were beginning to set up their song territories, females with feeders spent on average 23.5% of the day feeding at the feeder (range 5.2–41.7%) with on average 36.4% (range 6.2–75.0%) of their daily foraging time spent feeding on the provisioned food, so it formed a major source of their intake. The equivalent figures for use of feeders by males who had feeders in their territories was an average of 16.5% (range 4.8–34.6%) of the day spent feeding at the feeder and 48.6% (range 19.6–77.8%, $n = 16$) of the total daily foraging time spent on the provisioned food.

Effects of food on territory size

As we predicted, females with feeders had smaller territories than did control females (Table 4.5), but there was no effect on male territory size. This suggests that male territory defence is concerned with monopolising mates rather than food resources and supports the proposed 'two stage' process of settlement on breeding territories, with females first settling in relation to habitat and males then settling to monopolise the females.

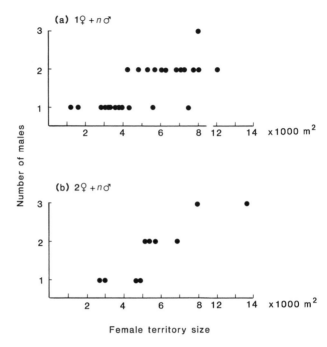

Fig. 4.8 The consequence of the variation in female territory size in Fig. 4.7 for the female's ease of monopolisation by males. The larger the female's territory the more likely she was to be defended by more than one male. (a) Plots of cases where one female associated with one male (monogamy) or two or three males (polyandry). (b) Plots of cases where two females associated with one male (polygyny) or two or three males (polygynandry). From Davies and Lundberg 1984.

Some birds were present in both years of the experiment and had feeders in one year but not the other. Four females had feeders in 1982 but not 1983 and six had feeders in 1983 but not in 1982. All ten females kept to almost exactly the same area of the Garden in both years and so matched comparisons could be done of the same bird on the same territory with or without a feeder. These matched comparisons showed the same result; the presence of feeders led to a decrease in female territory size (nine of the ten females had smaller territories in the year they had a feeder) but had no effect on male territory size (Table 4.5).

Effects on mating system

The effect of a feeder on the mating system was examined in two ways. First, we simply compared the frequencies of the various mating combinations in cases where the birds involved either had or did not have access to a feeder. In all cases of monogamy and polyandry, all the birds involved in the mating system

Table 4.5. Feeder experiment. The provision of extra food led to a decrease in female territory size but not male territory size (from Davies and Lundberg 1984). (a) Totals for the two years of the experiment. (b) Matched comparisons for individuals present on the same site in both years, who had a feeder in only one of the years.

	Territory size (m^2)						Significance of difference
	With feeder			Control			
	Mean	(SE)	n	Mean	(SE)	n	
(a) *Total*							
Females	2,776	(379)	28	4,572	(456)	39	$P < 0.01$
Males							
Territory defended by one male	2,864	(340)	11	2,642	(416)	13	NS
Territory defended by two males	5,276	(797)	14	6,614	(674)	17	NS
(b) *Matched comparisons*							
Females ($n = 10$)	2,841	(697)		7,249	(1,149)		$P < 0.01$
Males ($n = 7$)							
Territory defended by two males	5,473	(741)		7,933	(1,417)		NS

Significance levels in (a) tested by 2-tailed Mann-Whitney U-tests, in (b) by 2-tailed Wilcoxon matched pairs tests. Not sufficient data to test single male territories in (b).

had access to a feeder if one was present. However, this was not always true for polygynandry. For example, in some cases where two males defended two females the feeder was in the territory of one of the females, so that she and both males used it, but the other female did not. In other cases, the feeder was on the boundary of the two females so all individuals had access to the food. Even if only some of the females had access, their smaller territories were predicted to influence the ease with which her males could monopolise both her and other females within their territory. We therefore divided the mating combinations into those where at least one of the females plus her males had access to a feeder versus those where none of the birds had access. Table 4.6(a) shows that the effect of the feeders was to change the mating system away from polyandry towards polygynandry, in other words to increase male mating success.

Another way to analyse the data is to score the number of males associating with each female (Table 4.6b). The effect of the feeders was clear; with the reduction in female territory size the feeder females were more easily monopolised and so had fewer males defending them.

Table 4.6. Feeder experiment: Influence on the mating system (from Davies and Lundberg 1984).

(a) Mating combination	Feeder		Control	
	$n = 25$	(%)	$n = 33$	(%)
Unpaired ♂	0	(0.0)	4	(12.1)
Polyandry				
3♂ 1♀	0 ⎫	(24.0)	2 ⎫	(48.5)
2♂ 1♀	6 ⎭		14 ⎭	
Monogamy	11	(44.0)	10	(30.3)
Polygynandry				
3♂ 2♀	0 ⎫		1 ⎫	
2♂ 2♀	5 ⎪	(32.0)	2 ⎪	(9.1)
2♂ 3♀	2 ⎪		0 ⎪	
2♂ 4♀	1 ⎭		0 ⎭	

	Feeder		Control		
	$n = 37$ females		$n = 32$ females		
(b) Mean number of males defending a female (SE)	1.05	(0.07)	1.59	(0.10)	$P < 0.001$

(a) Comparing males who were unpaired plus males who shared one female (polyandry) versus males who had access to one (monogamy) or more females (polygynandry), the difference between feeder and control systems is significant, $\chi^2_1 = 6.29$, $P < 0.02$. (b) The number of males associating with each female is calculated by dividing the number of males by the number of females in the mating combination. Significance level tested by 2-tailed t-test.

Confounding factors

A possible confounding factor in the interpretation of the experiment is that of differences in bird 'quality' between feeder and control territories. It could be argued, for example, that the 'best' females settled where the feeders were located and these females were then likely to attract the 'best' males, namely those able to monopolise them most effectively. If this was true, then the mating system differences between feeder and control territories could have arisen due to differences in bird quality. This interpretation is unlikely because, as shown in Chapter 3, both males and females were sedentary and tended to occupy the same territories in different years; they did not move to follow the feeders. This meant that 'matched comparisons' could be made for birds present in both years, but which had extra food only in one of the years. Figure 4.9 shows an example.

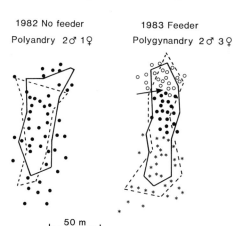

1982 No feeder   1983 Feeder

Polyandry 2♂ 1♀   Polygynandry 2♂ 3♀

50 m

Fig. 4.9 An example of the effect of a feeder. In 1982 no feeder was present and two males, whose song polygons are shown by solid (alpha) and dashed (beta) lines, shared one female whose range is indicated by (•). In 1983 a feeder was present, indicated by an arrow. The same two males defended approximately the same territory but the old female's territory contracted dramatically. As a consequence, two new females settled in the vacant area. One (○) shared the feeder with the old female and, like her, had a small territory. The other female (∗) did not have access to the feeder and her territory was considerably larger. The presence of the feeder thus changed the mating system from polyandry to polygynandry.

Female matched comparisons. Considering the ten females who were present in both years but who had extra food only in one (the ten females in Table 4.5b), Fig. 4.10 shows that their change in territory size between years was correlated with their change in mating system. Females whose territory increased from one year to the next had more males defending them while females whose territories decreased in size had fewer males. Thus changes in female territory size brought about changes in the mating system even considering the same females in different years. I regard this as the most convincing support for the hypothesis that female territory size influences a male's ability to monopolise mates.

Male matched comparisons. Now considering the same males on the same territories who had food in only one of the years, Fig. 4.11 shows that in all cases male mating success was as great or greater in the year with extra food. The analysis can be extended to include territories where there was food provided in only one year but different males were present in the two years. Again, mating success was greater in the year the territory had extra food. Furthermore, Fig. 4.11 shows that when territories are ranked in order of male mating success, there is a significant correlation between success in the year without food and the year with food. This means that the addition of a feeder to a territory with low male mating success did not increase male success to as

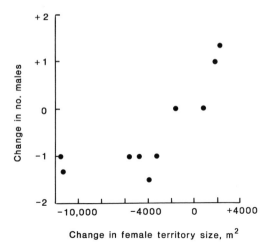

Fig. 4.10 For the 10 females which were present in both 1982 and 1983, but had a feeder in only one of these years, their change in territory size from 1982 to 1983 was correlated with their change in mating system from 1982 to 1983. A female had more males associated with her when her territory increased and fewer males when her territory decreased. Spearman rank correlation = 0.790, $P < 0.01$. Changes from ♂ ♀ to 2♂ ♀ are scored as a change in one male, from 2♂ ♀ to 2♂ 3♀ as a change from 2 to 0.67 males (i.e., a change of 1.33 males) and from 2♂ ♀ to 2♂ 4♀ as a change from 2 to 0.5 males (i.e., a change of 1.5 males). From Davies and Lundberg 1984.

high a level as did the addition of food to a territory with high mating success. The effect of a feeder, therefore, appears to add on to natural differences in territory quality already present.

Finally, we can ask whether the change in mating system brought about by the feeders is due simply to the decrease in female territory size, and hence their ease of monopolisation by males, or whether it is also due to changes in male behaviour. Males with feeders did not have larger territories (Table 4.5) but they may have had more spare time in which to attract and defend females. Table 4.7 compares the time budgets of males with and without feeders during the period 20 February–28 March 1983. Male time budgets taken at this time are likely to be most relevant to determining the mating system because this is the period during which most females settle on their breeding territories. Males with feeders began to sing earlier in the year, sang more, spent less time feeding, and more time perched and interacting with other dunnocks than did control males. There was, however, no significant variation in male time budget with mating system within either the feeder males or control males.[7] For example, within both the feeder and control territories males who sang more or perched more were not more likely to belong to one mating category rather than another.

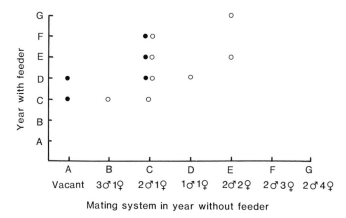

Fig. 4.11 Matched comparisons for the 13 male territories which had a feeder in one of the experimental years (1982, 1983) but not the other. The mating combinations are ranked in increasing order of male mating success. (○), the same male(s) were involved in both years or at least one of the males was the same. (●), territories where the male(s) were different in the two years. All 13 territories had as great, or greater, male mating success in the feeder year. Spearman rank correlation = 0.577, $P < 0.05$. From Davies and Lundberg 1984.

Because male time budget does not vary with mating system, the most likely explanation for the changes brought about by the feeders is that they are due simply to changes in female territory size.

4.6 Removal experiments to investigate male defence of females

The results so far suggest that the two sexes are playing different games. Females seem to ignore males altogether when setting up their territories, and simply settle in relation to food and nest sites. Males then compete to defend the females, with larger female territories being more difficult for a single male to monopolise. This idea was tested further by examining the effects of natural and experimental removals. If females are indeed indifferent to what the males are doing, then we would predict that if all the males were removed from the population then the females would carry on exactly as before, simply setting up their territories in relation to resource distribution. Obviously this extreme experiment is impractical, but male removals on a modest scale can test this prediction. Likewise, male removals can test whether larger female territories are indeed more difficult for single males to defend, as suggested by the fact that larger female territories more often lead to polyandry.

Table 4.7. Comparison of male time budgets in early spring 1983, for males with feeders and males without. Figures are means, with significant differences tested using Mann-Whitney U-test, 2-tailed (from Davies and Lundberg 1984).

Time budget	Feeder males ($n = 15$)	Control males ($n = 19$)	P
% time feeding	32.9	72.6	< 0.002
% time perched	47.8	17.5	< 0.002
% time preening	3.2	2.9	NS
% time interactions	16.1	7.0	< 0.02
Songs per hour	126.4	56.7	< 0.002
Interactions per hour	8.6	2.8	< 0.002
Mean date of first song (median) (1 = 17 January)	10.4 (5)	28.5 (32)	< 0.002

Influence of male removals on male mating success in polyandry and polygynandry

Sometimes, either the alpha or beta male in polyandry or polygynandry was killed by a predator, leaving a single male to defend the territory alone. Table 4.8 shows that in most cases the remaining male, whether alpha or beta, was successful in retaining the females. There was a hint that an alpha male was better able to retain two females than was a beta male, but the difference was not significant.

Ben Hatchwell and I supplemented these natural removals with experiments, where we caught a male and kept him in an aviary away from the territory for a three day period before then putting him back on his territory (Hatchwell and Davies 1992*b*). During the two years 1989 and 1990 we removed a total of 35 alpha males from polyandrous and polygynandrous mating systems. In 33 of these cases the beta male was successful in defending the territory alone during the alpha male's absence, and the system then reverted back to polyandry or polygynandry when the alpha male was put back. Only in two cases did another male settle on the territory during the removal, and in one of these he quickly left when the removed male was returned.

These results, together with those from the natural removals, show that a single male is capable of defending a territory that had previously been defended by two males. Admittedly, the experiments only tested ability to maintain the territory single handed during a short period of 3 days, but the natural removals showed that most single males could do so in the long term too. Further experi-

Table 4.8. The consequences of deaths from predation of either alpha or beta males in polyandry and polygynandry.

Mortality and consequence	No. cases when remaining male was:	
	Alpha male	Beta male
One male dies in polyandry, 2♂ 1♀		
Remaining male retains female, as monogamous pair	9	8
New male joins to re-form a trio		
New male becomes alpha	0	1
New male becomes beta	2	2
One male dies in polygynandry, 2♂ 2♀		
Remaining male retains both females, as polygyny	3	3
New male joins to re-form polygynandry		
New male becomes alpha	0	0
New male becomes beta	0	4
Total cases retained female/females*	12/14	11/18

*Difference between remaining alpha and beta males is not significant, $\chi^2_1 = 1.298$, $P > 0.20$.

ments by Philip Byle, who removed five alpha males and five beta males from polyandrous trios for periods of 3 to 14 days, likewise showed that a single male could usually maintain the territory in the other's absence; in only one of these cases did a newcomer settle during the removal period (Byle 1987).

In Byle's (1987) study, although the single male retained the female, there was a 25% decrease in his territory size during the removal period. In the experiments by Ben Hatchwell and myself, however, there was no such decrease (Fig. 4.12). Even in Byle's experiments, territory size did not decrease by half, which is what might be predicted from the fact that two males, in polyandry and polygynandry, defend territories twice the average size of monogamous males (Table 4.1).

Although the effects of removals on territory size are conflicting, there is no strong support, even from Byle's experiments, that polyandry arises because two males are needed for successful defence of the female territory against neighbours (which has been suggested for pukekos *Porphyrio porphyrio*, Craig 1984). However, in all cases the removed males were able to settle back on their territories when we released them (Hatchwell and Davies 1992*b*). This shows that although a single male could keep neighbours at bay, he could not

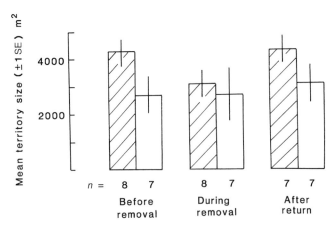

Fig. 4.12 Influence of number of males defending a territory on male territory size. In experiments by Byle (1987) removal of one polyandrous male led to a decrease in the remaining male's territory size (hatched columns, Wilcoxon matched pairs test, $P < 0.01$), and the size increased back to its original value when the removed male was put back on the territory. Removal experiments by Hatchwell and Davies (1992*b*) (open columns) showed no such effect. When alpha males were removed from polyandry and polygynandry, the remaining beta male maintained the same territory size as before and likewise showed no change when the alpha male was put back (Friedman 2-way ANOVA, $\chi^2_2 = 0.86$, NS).

evict a persistent intruder. At the time of our experiments (April–June), all males had settled on breeding territories. If we had done experiments earlier (February–March), when unpaired males were roaming the Garden, then our removals may have caused increased settlement by newcomers. It is possible that polyandry arises in association with large female territories (Table 4.1) partly because unpaired males find it easier to settle on larger territories.

Influence of male removals on females

There was no effect of removal of one of the polyandrous or polygynandrous males on female territory size. Likewise, when we removed a monogamous male, female territory size remained the same (Fig. 4.13). When second males joined a monogamous pair to form a polyandrous trio, there was also no change in either male or female territory size (Fig. 4.14).

These results provide strong support for the idea that female territory size is completely independent of male influence and, as shown earlier, determined only by competition with other females for suitable habitat. Female territories simply form a template upon which male territories are superimposed.

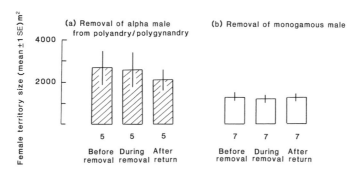

Fig. 4.13 (a) Removal and subsequent return of an alpha male from polyandry and polygynandry (hatched columns) had no effect on female territory size (Friedman 2-way ANOVA, $\chi^2_2 = 0.40$, NS), nor did (b) removal and subsequent return of a monogamous male (open columns) ($\chi^2_2 = 4.57$, NS). From Hatchwell and Davies 1992*b*.

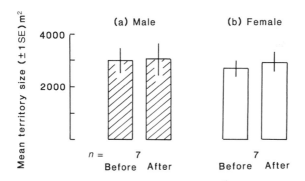

Fig. 4.14 When a second male settled on a monogamous pair's territory, thus changing the mating system to polyandry, there was no change in either male (hatched columns) or female (open columns) territory size. In this analysis I have excluded cases where a male arrived due to the death of a neighbouring female because changes in territory size may then have been confounded by the female vacancy.

4.7 Summary

The variable mating system was caused by variation in both male and female territory size. Single males with larger territories were more likely to be polygynous and territories defended by two males were more likely to contain several females when female territory size was small.

The mating systems formed by females first setting up exclusive territories in relation to resources and males then competing to monopolise the female territories. Polyandry and polygynandry occurred when one male was unable to defend the female(s) alone, either by coalescence of neighbouring male territories or by settlement of an unpaired male on the territory.

The size and location of female territories was completely independent of male distribution and influenced only by competition with other females. Thus, when females were removed by predation, neighbouring females often moved or expanded to fill the vacancy. By contrast, when males were removed there was no effect on female territory size or location.

Males moved to settle on the female distribution, not vice versa. When the death of a female left a male without a mate, he moved to compete for another female. By contrast, when the death of a male left a female without a mate, she remained and other males moved in to defend her. Males were quick to respond to vacancies and moved to increase their mating success.

Female territories were smaller in good feeding areas with dense vegetation. The larger a female's territory, the more likely she was to be defended by more than one male. When food was provided, female territories decreased in size and male mating success increased. Thus, females influenced the mating system indirectly, by setting up territories in relation to resource distribution, with the size of their territories then influencing their ease of monopolisation by males.

STATISTICAL ANALYSIS

1. Data on territory size were collected in the seven years 1981–1984 and 1988–1990. Territories in 1982 and 1983 which were involved in a feeder experiment are excluded in this analysis. Kruskal-Wallis 1-way ANOVAs showed no significant variation between years in either male or female territory size for any of the mating systems.

2. Comparing alpha and beta male territories in 19 cases of polyandry, mean territory sizes were alpha 5,077m^2 (SE = 627), beta 4,906m^2 (SE = 629). Wilcoxon matched pairs test, $T = 81$, NS. In 10 cases the alpha male polygon was larger, in 9 cases the beta male polygon was larger.

3. Comparing alpha and beta male territories in 25 cases of polygynandry, mean territory sizes were alpha 5,019m^2 (SE = 347) and beta 4,294m^2 (SE = 345). Wilcoxon matched pairs test, $T = 34$, $P < 0.01$, 2-tailed. In 16 cases the alpha male polygon was larger, in 7 cases the beta male's polygon was larger and in 2 cases they were of equal size.

4. Spearman rank correlation between initial female territory size and size increase after occupation of the vacancy, $r_s = 0.602$, $n = 16$, $P < 0.01$.

5. Spearman rank correlation, $r_s = 0.684$, $n = 12$, $P < 0.05$.

6. Differences in female territory size with habitat type were consistent between years. For 8 of the sites in Fig. 4.6 where there were data on female territory size for 6 years, there was consistent ranking of territory size between years; Friedman 2-way ANOVA, $\chi^2_7 = 28.014$, $P < 0.001$. In this analysis, some females counted more than once

because they bred in several years; if each female is scored only once, in her first year, the difference is still highly significant, $\chi^2_7 = 26.819$, $P < 0.001$.

7. Kruskal-Wallis 1-way analysis of variance comparing time budgets of unpaired, polyandrous, monogamous and polygynandrous males within the feeder males and within the non-feeder males; no significant differences.

5

Factors influencing an individual's competitive success

5.1 Body size and age

I now turn to the question of what might influence a female's success in competing for habitat and a male's success in competing for mates, particularly the effects of body size, age, and familiarity with a territory.

Wing length was measured from the carpel joint to the tip of the longest primary, with the wing closed and pressed flat against a rule held along the long axis of the bird's body (Fig. 5.1a). Comparing all individuals measured at all ages, males had longer wings than females and in both sexes wing length increased from year 1 to year 2 but thereafter there was no further change (Fig. 5.2a). Measurements of individual changes with age supported this conclusion, with males and females who were measured in both year 1 and year 2 showing significant increases in wing length, but individuals measured in years 2 and after showing no change.[1]

Tarsus lengths were measured, with vernier callipers, from the notch in the angle of the intertarsal joint to the tip of the bended foot (Fig. 5.1b). This measure is slightly larger than the true tarsus (Svensson 1970) but it is easier to take and more repeatable. Males had longer tarsi than females, but neither sex showed any changes with age (Fig. 5.2b).

5.2 Body size and mating system

Males and females involved in different mating systems did not differ in tarsus length, nor did they differ in wing length (considering same-aged individuals, to control for the age change in wing length shown above).[2] Comparison of alpha versus beta male measurements within the same mating combination likewise revealed no differences in tarsus length, but alpha males had signifi-

(a)

(b)

Fig. 5.1 Measuring (a) wing length and (b) tarsus length.

cantly longer wings than beta males in both polyandry and polygynandry (Table
5.1). Given the lack of difference in wing length between alpha and beta males
of the same age this must simply reflect the differences in age between alpha
and beta males, shown below.

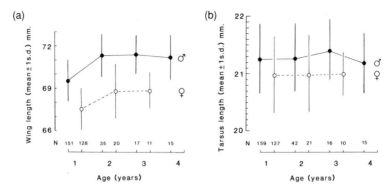

Fig. 5.2 Body size measurements. (a) Both male ($P < 0.001$) and female ($P < 0.01$) wing length increased from year 1 to year 2, but there was no further change. (b) There was no change in tarsus length with age for either sex. Males had longer wings ($P < 0.001$) and longer tarsi ($P < 0.001$) than females. Significance levels refer to 2-tailed *t*-tests. Sample sizes given above the *x*-axis.

5.3 Age and mating system

(a) *Comparison across all individuals*. Table 5.2 summarises the ages of males and females in the different mating systems. It shows that there were no differences among the females in the different systems but marked differences among the males. Male mating success increased with age. Beta males in polyandry and polygynandry were mostly first-year birds. Comparing the other cases there was also a clear trend for male success to increase with age, with a greater proportion of older males in systems with more than one mate (polygyny and alpha polygynandrous males).

Comparing 63 cases of polyandry, the alpha male was the older male in 33 cases, the same age as the beta in 27 cases and the younger in only 3 cases.[3] Similarly, of 60 cases of polygynandry, the alpha male was the older male in 40 cases, the same age as the beta male in 13 cases, and the younger of the two in only 7 cases.[4] Although the older of the males was more likely to be alpha in both systems, there was a significant tendency for very old males to drop to beta status (Table 5.3). The data show that compared to 3-year-olds, males aged 4 or more were equally likely to dominate first-year males but they were less likely to be alpha in conjunction with a 2-year-old.

Thus males clearly increase in status as they get older but in the very old age classes there are significant signs of decreased mating success. Two of the three males who reached the grand old age of 7 or more remained unpaired throughout the breeding season, occupied small territories in between other breeding territories and mainly fed quietly alone, showing little competitive spirit!

Table 5.1. Comparisons of alpha and beta male measurements within the same mating systems, for polyandry and polygynandry. Data given are means, with 1 SE in brackets.

Mating system	Alpha male	Beta male	Significance of difference
Tarsus length (mm)			
Polyandry (*n* = 61)	21.34 (0.08)	21.21 (0.08)	NS
Polygynandry (*n* = 51)	21.37 (0.08)	21.15 (0.08)	NS
Wing length (mm)			
Polyandry (*n* = 57)	70.75 (0.19)	69.83 (0.21)	*P* < 0.01
Polygynandry (*n* = 44)	71.11 (0.22)	69.89 (0.27)	*P* < 0.001

Significance levels refer to 2-tailed Wilcoxon matched pairs tests.

(b) *Individual changes with age*. These analyses of age changes involve comparison of different individuals and so raise a problem. If poorer quality individuals were less likely to survive, then there would obviously be a greater proportion of high quality individuals in the older age classes. The apparent age differences in male success may then merely reflect differences in individual quality. To control for this possibility, I examined whether particular individuals also changed in mating success with age. In Table 5.4 I have scored whether particular individuals increased, decreased or remained the same with respect to their access to mates as they got older. For females I have ranked the mating systems in the following order of increased access to males; polygyny (share one male), polygynandry (share two or more males), monogamy (one male), polyandry (two males). Just as with the 'across individual' analysis, there was no sign of any changes in female mating status with age.

For males, however, Table 5.4 shows the same strong age effects as before. Male mating status was ranked in the following order of increased access to females; unpaired, beta male polyandry, beta male polygynandry, alpha male polyandry, monogamy, alpha male polygynandry, polygyny. I shall show in Chapter 9 that this corresponds to increasing male reproductive success. From 1 to 2 years of age, individual males were likely to increase in success but thereafter there was a greater probability of a decrease.

This 'within individual' analysis thus supports the conclusions from the 'across individual analysis', namely the absence of age effects on mating system

Table 5.2. Ages of males and females involved in the various mating systems. The total column includes some birds known to be older than one year, but whose exact age was not known.

Mating system	No. individuals of each age (only birds whose exact age was known)						Total		
	1yr	2yr	3yr	4yr	5yr	6yr	1yr	> 1yr	(% > 1yr)
(a) *Females*									
Polyandry	37	15	3	2	1	–	37	24	(39.3)
Polygynandry	70	26	9	3	3	–	70	51	(42.1)
Monogamy	41	24	9	1	2	–	41	38	(48.1)
Polygyny	7	4	4	1	–	–	7	10	(58.8)
(b) *Males*									
Polyandry β	48	8	–	3	–	–	48	15	(23.8)
α	20	9	6	2	4	1	20	43	(68.2)
Polygynandry β	40	6	2	1	–	1	40	18	(31.0)
α	9	18	13	5	–	1	9	49	(84.5)
Monogamy	33	25	10	4	1	1	33	50	(60.2)
Polygyny	2	3	1	1	1	–	2	7	(77.8)

(a) *Females*. No significant variation with age: total column, $\chi^2_3 = 2.75$, $P > 0.30$.
(b) *Males*. Highly significant variation with age: total column, $\chi^2_5 = 65.69$, $P < 0.001$. Analysing further with χ^2 tests (1df), beta males significantly younger than alpha males in both polyandry ($P < 0.001$) and polygynandry ($P < 0.001$). No difference in beta male age in polyandry and polygynandry, but alpha males significantly older in polygynandry than polyandry ($P < 0.05$) and also than monogamous males ($P < 0.01$). No difference between alpha males in polyandry and monogamous males.

in females and the tendency for males to increase in mating success with age until they reach 3 or more, when there is evidence of senescence.

Causes of age changes

Why do male dunnocks show such a marked improvement in mating success from their first to their second years, and to a lesser extent from ages two to three? Age changes in mating and reproductive success have been well documented in seabirds (e.g., Coulson 1966; Ollason and Dunnet 1978; Wooller *et al.* 1990) and raptors (Newton 1988) but they are evident even in short-lived passerines (e.g., Perrins and McCleery 1985). Do young birds do less well than older birds because of 'restraint' (they do not try so hard) or 'constraint' (they are not able to do so well; Curio 1983)?

Table 5.3. Ages of alpha and beta males in polyandry and polygynandry. Alpha males are usually older than beta males though very old males (> 4 years old) are more likely to be beta males when they associate with other males of 2 or more years of age.

Age of older male (yrs)	No. cases older male is alpha	
	Other male 1 year old	Other male 2 or more years old
2 or 2+	31/31	–
3 or 3+	13/15	7/7
4+	17/17 *	5/13†

* Males of 4+ years old are significantly more likely to be alpha male when they associate with a one year old male than when they associate with a male of two or more years of age, Fisher exact test *P* < 0.001. † Males of > 4 years old are less likely to dominate a male > 2 years old than are males 3 or > 3 (Fisher exact test *P* < 0.05).

Detailed studies of blackbirds *Turdus merula* in the Garden, by André Desrochers and Robert Magrath, have shown that there are marked improvements in foraging ability with age and that these are likely to be the main cause of age differences in reproductive performance in this species (Desrochers in press). Unfortunately, I was unable to measure foraging success in dunnocks because they fed on such tiny prey. However, it is possible that the age changes in male success were likewise linked to improvement in foraging, perhaps due to increased familiarity with a territory (e.g., knowledge of feeding sites and potential dangers), which would allow individuals more time to devote to competition for mates. Greater familiarity with a territory may explain why alpha males were usually the older of the two males in polyandry and polygynandry. When alpha males died, beta males then, in turn, often became alpha to a new first-year bird.

Table 5.5 shows that the increase in male success from years one to two was due both to an increase in number of females and increased success at keeping other males at bay. Thus several polyandrous males became polygynandrous (an increase in number of mates) and several beta males became alpha males or monogamous males (an increase in success at dominating or evicting male competitors). There was no evidence for any age change in male territory size,[5] so the increase in number of mates must have been caused by gaining control of areas where females were easier to defend.

By contrast, female territory size decreased with age.[6] Although the cause of this is not known, this may link up rather neatly with the different functions

Table 5.4. Individual changes in mating status with age. There was no effect among females but males were increasingly likely to have decreased mating status as they got older. Mating status was measured as access to mates, including number of mates and (for males) dominance rank (see text).

Age change (yrs)	Changes in an individual's mating status							
	No. males				No. females			
	Increased	Remained same	Decreased	(% decreased)	Increased	Remained same	Decreased	(% decreased)
1–2	41	20	10	(14.1)	15	35	25	(33.3)
2–3	18	9	7	(20.6)	5	14	14	(42.4)
3–4	6	6	13	(52.0)	3	2	3	(37.5)
4–5	3	4	3	⎫	4	–	2	(33.3)
5–6	–	3	3	⎬	–	–	–	
6–7	1	–	1	⎭ (42.1)	–	–	–	
7–8	–	–	1		–	–	–	

Comparing the frequencies of decrease in mating status, males were more likely to decrease as they aged ($\chi^2_3 = 17.19$, $P < 0.001$) but there was no significant effect among females ($\chi^2_2 = 2.56$, $P > 0.20$).

Table 5.5. Changes in the mating status of individual males from their first to their second breeding seasons. Mating status is arranged in increasing order of access to mates (see text). The table gives the number of individuals who made each transition.

Mating system in year 1		Mating system in year 2						
		U	PA_β	PGA_β	PA_α	M	PGA_α	PG
Unpaired male	(U)	–	–	1	–	–	–	–
Beta male in polyandry	(PA_β)	1	4	3	3	8	1	1
Beta male in polygynandry	(PGA_β)	–	2	4	3	3	2	2
Alpha male in polyandry	(PA_α)	–	–	–	2	3	6	–
Monogamy	(M)	–	–	2	–	8	5	–
Alpha male in polygynandry	(PGA_α)	–	–	–	1	2	2	–
Polygyny	(PG)	–	–	–	–	1	1	–

of male and female territories shown in Chapter 4. If female territoriality is mainly concerned with competition for resources, especially food, then increased familiarity with an area may lead to more efficient resource exploitation and allow a decrease in territory size. Alternatively, females may change their territory boundaries to incorporate better sites, again allowing a decrease in size. Males, on the other hand, defend territories to monopolise mates and no such decrease with age would be expected.

5.4 Male status and territory value: remove and release experiments

Many of the age changes shown above may have come about as a result of increased familiarity with a territory, mates and competitors. The influence of familiarity on competitive success can be tested by experiments where a territory owner is removed for a short period, thus allowing a newcomer to take over the vacancy. The original owner is then released back on his territory. Such experiments with willow warblers *Phylloscopus trochilus* have shown that original owners are more likely to win their territory back if they are older and have had greater experience of the territory (Jakobsson 1988). In great tits *Parus major* and red-winged blackbirds *Agelaius phoeniceus* original owners are less likely to win their territory back the longer the newcomer has been in residence (Krebs 1982; Beletsky and Orians 1987). Thus an individual's familiarity with a territory seems to influence competitive success.

In 1989 and 1990, Ben Hatchwell and I performed similar 'remove and release' experiments (Hatchwell and Davies 1992*b*). The main aim of these was to investigate the influence of paternity on parental effort (Chapter 12), but we incidently obtained some interesting results which showed how experience with a territory could determine the outcome of contests and alpha versus beta status among males. We removed a total of 52 males, 26 monogamous males and 26 alpha males, from polyandry and polygynandry. The males were caught in mist nets and kept in aviaries about 1km away from the Garden for 2 to 9 days (most cases 3 days) and then released back on their territories. All these males retained good body weight and health during their short period in captivity and they all quickly settled back on their original territories when released. In one case, on the morning a male was due to be returned to the wild, we opened the cage door rather carelessly and the male quickly shot out to freedom. Thinking the experiment had been ruined, we sadly cycled the kilometre back to the Garden only to discover that the male had beaten us, and was already singing back on his old territory! This was a gentle reminder that the birds usually knew far more about what was going on than we did.

All the removals were done during the breeding season on territories where the females had a complete nest but they were done at different stages in relation to egg laying, with some males removed before the laying of the first egg and others after egg laying had begun.

(a) *Monogamous males.* When monogamous males were removed, neighbours took over the territory. Either a neighbouring monogamous male expanded to become polygynous or neighbouring polyandrous or polygynandrous males took over the temporarily widowed female. When the original owners were released back on to their territories, 16 were successful in driving off the newcomer and so regained their former status as a monogamous male, while the other 10 ended up sharing the territory with the newcomer. Two original owners became alpha males in polygynandry, seven became beta males and one became a gamma male.

Table 5.6 divides the original owners into cases where they retained status, either by driving out newcomers altogether or at least dominating them as an alpha male, and cases where they lost status to the newcomers, becoming beta or gamma males. The results show that the original owner was more likely to lose status if he was removed before egg laying, and so the newcomer had a chance to copulate with the female during the key time for fertilising eggs (see Chapter 6). Once egg laying had begun, the original owner could be sure that

Table 5.6. Results of 'remove and release' experiments to investigate factors influencing competitive success in males (from Hatchwell and Davies 1992*b*).

Original owner removed	Males who took over	No. original owners who lost status when released	
		Removed before laying	Removed after laying
Monogamous male	Neighbour	7/14 (50%)	1/12 (8%) $G_1 = 5.80$ $P < 0.02$
Alpha male in polyandry and polygynandry	Resident beta	7/8 (88%)	8/18 (44%) $G_1 = 4.67$ $P < 0.05$

A higher proportion of alpha males lost status than did monogamous males ($G_1 = 3.87$, $P < 0.05$), a trend evident both before laying ($G_1 = 3.41$, $P < 0.10$) and after laying ($G_1 = 5.04$, $P < 0.025$).

he had fathered some of the clutch, and he was much more likely to retain status. The duration of the removal period had no influence at all on loss of status, only the duration of the period in which the newcomer associated with the female during her egg laying period (Hatchwell and Davies 1992*b*).

The outcome of contests were therefore related to how the competitors valued the territory in terms of potential reproductive success. Original owners who had gained increased mating access were probably prepared to fight harder to get their territories back, while at the same time newcomers who had gained less mating access may have been less prepared to contest for dominance.

There was no influence of body size (measured by tarsus length) on the outcome of contests between original owners and newcomers. However, although original owners who lost status were no different in wing length from those who retained status, newcomers with longer wings were more likely to usurp the territory. This result is certainly a reflection of age (see Section 5.1); there was no difference in the age of original owners who lost or retained status but newcomers who were successful usurpers were significantly older than those who were not (Hatchwell and Davies 1992*b*). These experimental results therefore support the conclusion from Section 5.3 that a male's competitive ability increases with age.

(b) *Alpha males*. When alpha males were removed, the resident beta male took over charge of the territory and in almost all cases was successful in keeping other competitors at bay (Chapter 4). Table 5.6 shows that, just as with monogamous male removals, the stage of removal in relation to laying had a

Male dominance status on a territory can be changed experimentally by short-term removals. When a removed male is returned, there is often a fight with the replacement male. The replacement is more likely to gain control if the removal is done during egg laying, which gives him the chance to gain paternity and so increases the value of the territory to him.

marked effect on whether the original male retained alpha status. Males removed before the onset of laying, who therefore missed part or all of the key period for fertilising eggs, were more likely to lose status. There was no influence of body size or age on the outcome of the contests (Hatchwell and Davies 1992*b*).

The dominance status of the two males seemed to be settled quite quickly. When the original owner was released back on his territory he usually flew up into a tree to sing. The other male then came over and chases occurred. In some cases the two birds grappled briefly on the ground and were so intent on their struggle that I was able to approach to within a metre or so, and in one case even touched them. Such intense contests also occur in other cases where territory ownership is confused by experiments (Davies 1978; Krebs 1982). However, after an hour or so chasing and singing, the dunnocks came to a stable dominance order with one being a clear alpha and able to displace the other from the vicinity of the female or from feeding sites.

Table 5.6 shows an interesting difference between the effects of removals of monogamous and alpha males. Alpha males were much more likely to lose status to their resident beta, than were monogamous males to newcomers. This was not because alpha males were younger than monogamous males; in fact there was a tendency for them to be older, as expected from the older average age of alpha polygynandrous males compared to monogamous males (Table 5.2). Likewise, the usurping beta males tended to be younger than usurping neighbours, again as expected from the fact that beta males are most often first-year males (Table 5.2). The greater tendency for removed alpha males to lose status, therefore, has nothing to do with age effects. The most likely explanation is that the alpha males, on release, were contesting for dominance with another male who was also a previous resident and so familiar with the territory, its female and the neighbours. By contrast, monogamous males on release were contesting with newcomers who had just recently occupied the territory and so had less advantage of familiarity.

In conclusion, these experiments reveal two factors influencing the outcome of contests between males. The first is the value of the territory, measured by a male's potential reproductive success. This suggests that there may be a 'self-reinforcing effect' of alpha male status; alpha males are better able to monopolise mating access (Chapter 6) and, as a result, the territory becomes more valuable to them so they are prepared to fight harder and, in turn, are then more likely to retain alpha status. This may be similar to the 'success breeds success' effect, well known to boxing trainers and football managers. Secondly, the experiments support the idea, suggested earlier in the chapter, that the age effects on male mating success may, in large part, reflect the advantages of increased familiarity with a territory.

5.5 Summary

Males had longer tarsi and wings than females. An individual's wing length increased from year 1 to year 2, but not thereafter. Males and females in different mating systems did not differ in tarsus length or wing length (controlling for age).

Females in different mating systems did not differ in age. There were, however, marked age differences in males. Beta males in polyandry and polygynandry were mostly first-year birds, and there was a greater proportion of older males in mating systems with more than one female (alpha males in polygynandry and polygynous males). Changes in individual male success with age supported these conclusions; as a male got older he was better able to dominate or evict male competitors from his territory and better able to monopolise more than one female.

There was evidence for senescence in males. Beyond the age of 3 years, males became increasingly likely to decrease in mating success. Compared with 2–3-year-olds, 4-year-old males in polyandry and polygynandry were equally able to dominate 1-year-olds but less able to dominate a 2-year-old.

Males who were removed from their territories for about 3 days, and then returned, were more likely to lose dominance to a newcomer who had gained access to the female during the egg laying period. Thus contest outcome was influenced by territory 'value', measured by potential male reproductive success. In addition, original owners were more likely to lose against older newcomers, and against opponents who were more familiar with the territory.

The age effects on male mating success may, to a large extent, reflect the advantages of increased familiarity with a territory.

STATISTICAL ANALYSIS

1. *Males*: Wing length of 31 individuals measured at both 1 year of age (mean 69.90, SE 0.26) and 2 years of age (mean 71.42, SE 0.29), $t = 3.913$, $P < 0.001$, 2-tailed. Comparing 18 individuals measured at 2 years (mean 71.11, SE 0.29) and 3 or older (mean 71.38, SE 0.36) there was no significant difference, $t = 0.574$, NS.

Females: Comparing wing lengths of 25 individuals measured in year 1 (mean 67.92, SE 0.35) and year 2 (mean 68.96, SE 0.39), $t = 1.993$, $P < 0.10$, 2-tailed. Of these, 19 increased in wing length, 4 remained the same and 2 decreased (sign test, $P < 0.01$, 2-tailed). Insufficient data to test for changes beyond year 2.

2. One-way ANOVAs all non-significant.

3. Alpha males in polyandry were more often the older of the two males, sign test, $P = 0.009$, 2-tailed.

4. Alpha males in polygynandry were also more often the older of the two males, sign test, $P < 0.001$, 2-tailed.

5. Territories defended by one male; no difference in territory size (m^2) between 1-year-olds (mean $\pm$ 1 SE, 2,735 $\pm$ 307, $n = 21$) and 2-year-olds or older (3,098 $\pm$ 268, $n = 33$), $t_{53} = 0.889$, NS. Territories defended by two males; also no difference between 1-year-olds (4,877 $\pm$ 623, $n = 15$) and 2-year-olds or older (5,558 $\pm$ 341, $n = 43$), $t_{57} = 0.958$, NS.

6. Territories defended by 1-year-old females were larger in size (m^2) (3,112 $\pm$ 245, $n = 81$) than those of birds 2 years old or older (2,547 $\pm$ 196, $n = 61$), $t_{141} = 1.801$, $P < 0.10$, 2-tailed. Considering 29 individuals whose territories were measured at 1 year old (3,960 $\pm$ 554) and at 2 or more years old (2,702 $\pm$ 349), territory size also decreased with age—Wilcoxon matched pairs test, $P = 0.013$.

6

Mate guarding and mating: sexual conflict

6.1 Breeding conflict begins

At the end of Chapter 4, we were left with two scenarios as mating combinations formed at the start of the breeding season. One was of some males coming to an amicable agreement to share the defence of a territory with a subordinate beta male. The other was of females playing no active part in determining the mating system and apparently ignoring males altogether when they set up their breeding territories in relation to food distribution and nest sites.

When breeding begins, however, both pictures are shattered. The alliance between the males who share a territory becomes an uneasy one; although they continue to cooperate in defence against neighbours and intruders, once their females begin to solicit copulations the two males compete intensely to gain mating access. For their part, the females no longer remain as passive participants but play an active role in determining which males copulate. A visit to a dunnock territory during the mating period presents all the drama and excitement of an avian soap opera.

6.2 The breeding cycle

From the end of March and throughout April, females begin to build nests in their breeding territories. All nests were built by the female alone, in hedges or evergreen shrubs, usually within 1.5m of the ground. The foundation was of small twigs, followed by a middle layer of grass stalks, bark, dead leaves and moss, then a nest cup consisting largely of green moss, and finally a thin, neat lining of hair or, occasionally, feathers (Tomek 1980). In the Garden the most popular source of lining was hair from squirrels and rabbits but some females would fly across several territories to collect hair from an enclosure containing the Director's pet donkey!

Females first spent several days prospecting for sites, hopping in and out of dense bushes and hedges, often accompanied by a male who followed quietly behind. Once a site had been chosen the female worked away alone, bringing material every few minutes, usually collected from the lawns or undergrowth within a distance of 20m or so. The males occasionally sang near by but mainly left to feed in other parts of the territory. The first nests of the season often took a week or more to build because females gave up during periods of cold weather and so, with a lengthy building period, they were easy to find during their construction. Later in the season, however, females completed new nests within a day or two and it was a hard job to keep track of events, especially when nests were depredated and replacements were quickly built.

The eggs, like those of other species of accentor, are a beautiful uniform blue. Most females had two or three broods per year, usually laying 3−5 eggs per clutch (Davies and Lundberg 1985). Normally, one egg was laid each day, early in the morning, until the clutch was complete, with incubation beginning on the day the last egg was laid. However, in cold weather, especially early in the season, females sometimes missed a day between laying some of the eggs. Incubation was entirely by the female and usually lasted 11−12 days, but in cold weather this could increase up to 17 days. Two females, whose eggs all failed to develop, incubated for 19 days before abandoning their clutches. Brooding of small chicks was done by the females alone, but both males and females fed the young, for a period of 11−12 days in the nest and then for another 2−3 weeks after fledging, until they reached independence.

Laying of new clutches came to an end in early July. For successful breeding attempts, the whole cycle took on average 43 days, so only females who began by early April had the chance to fit in three successful broods in a season and most had time for only two.

6.3 Mate guarding behaviour

In many animals, males guard their mates during the fertile period to prevent other males from copulating. Such 'mate guarding' protects the male's paternity and was first described in insects (Parker 1970). Since the work of Tim Birkhead (1979, 1982) on magpies *Pica pica* and Michael Beecher (M.D. Beecher and I.M. Beecher 1979) on bank swallows (= sand martin) *Riparia riparia*, it has been recognised as being widespread in birds too. During the mating period I spent long hours sitting quietly watching female dunnocks and simply recording in a notebook which males associated with them, how close they were, how often chases occurred and the frequency of copulations. I tried to watch some females for an hour or two every day until incubation began, to get an idea of how the intensity of competition for matings varied with time and with the mating system.

Mate guarding. TOP, *the alpha male, perched on a post, keeps close watch while the female (left) feeds. The beta male (behind) remains close by, awaiting an opportunity to mate.* BOTTOM, *the female flies off with both males in hot pursuit. Females often attempt to escape the alpha male's close attentions and encourage matings by the beta male.*

Once the nest cup is complete, and at about the time she begins to line the nest, the female dunnock starts to solicit copulations with a characteristic display (see later). The males then begin to associate closely with the females, and the games begin! Monogamous males chased off neighbouring males who trespassed in search of extra matings, but by far the most intense competition occurred when two (or more) males shared a territory. Here, the alpha male guarded the female

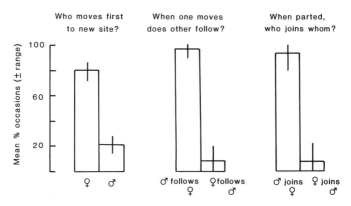

Fig. 6.1 Mate guarding: males follow females, not vice versa. Histograms show means and ranges from seven territories for guarding by monogamous males ($n = 3$) and alpha males ($n = 4$).

closely, following her wherever she went, and attempted to prevent the beta male from gaining access to her. Subordinate males continuously approached the female throughout the mating period and alpha males sometimes spent 40 minutes per hour chasing them off. There were occasional physical clashes between the males when they grappled briefly on the ground before continuing their 'cat and mouse' chases through the vegetation in the female's vicinity.

During mate guarding it was clear that the male was responsible for maintaining the close association with the female. The female initiated most of the flights to new feeding areas and the male nearly always followed her. On the few occasions the male flew off first, the female rarely followed him. Furthermore, when the two were parted it was the male who joined the female rather than the reverse (Fig. 6.1).

Alpha males sometimes lost the female and the beta males were then able to copulate in peace if they found her first. The alpha male would fly round the territory searching frantically for the female, perching occasionally to sing or look about, and then flying off again, making short 'durrup' calls as he looked amongst all the bushes and rockeries. The female would make no effort to rejoin him and often remained hidden, feeding in dense undergrowth, sometimes with the beta male, until the alpha male found her, whereupon the guarding and chasing began all over again. Alpha males often lost the female for periods of up to 15 minutes and once I watched one search for 3 hours before he found her. It was remarkable how the female could hide on a territory whose features must have been so familiar to the searching male.

It was common for the alpha male to lose the female after chasing the beta male away from her vicinity, and the continuous probing by the beta male often seemed to be an attempt to lure the alpha male into starting a chase, perhaps to increase the chance that he would lose her as a result. Sometimes, both males lost the female, in which case the alpha usually followed the beta male around the territory until she was found again. It made good sense for the alpha male to do this because if the beta male was left to search alone, he would have gained the chance of mating with the female if he was the first to find her.

6.4 Intensity of mate guarding

Comparison of monogamous and alpha males

In monogamous pairs the male and female were rarely disturbed by neighbours and copulation usually proceeded in peace. Where two males shared a territory, however, the males frequently interrupted each other's copulation attempts. Figure 6.2 shows that the probability of the other male interrupting decreased with his distance away, following the same curve for both alpha and beta male. The decrease occurred simply because the lone male was less likely to see the start of a mating display when he was further away. The graph shows that the alpha male had to be within 5m of his female to be sure of preventing the beta male from copulating successfully.

With the constant threat posed by a resident beta male, it was not surprising that alpha males in polyandry spent more time very close to the female ($<$ 5m) than did monogamous males (Fig. 6.3a). Some matched comparisons could be done of the same male guarding as a monogamous male and as an alpha male in polyandry because some pairs were joined by beta males for later mating attempts and, in other cases, a trio became a pair when one of the males died. Figure 6.3b shows that a male spent more time within 5m of the female when there was a beta male present on his territory.

This suggests that there must be a cost for close guarding, otherwise monogamous males would guard as closely too. One cost may be time spent by the male following the female, time which he could presumably have spent feeding or resting. However, close guarding was also costly for females because it reduced their feeding rate when they were foraging on small invertebrates by hopping rapidly over the ground and picking up surface prey (Fig. 6.4a). Another bird close by may have disturbed the prey or have eaten items the female herself would have collected. There was no such cost of close guarding when the birds fed on grass seeds (Fig. 6.4b). Here, the dunnocks often remained in one area picking up seeds from small patches, so interference was apparently less of a problem.

Mate guarding also began significantly earlier in relation to egg laying when there was a resident beta male on the territory (Davies 1985). An alternative explanation of these data is that guarding did not start earlier but rather caused a delay in the onset of laying when two males competed for matings, perhaps because of the adverse effect of close guarding and chasing on the female's

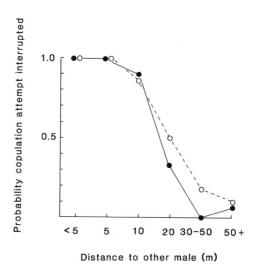

Fig. 6.2 Probability that a copulation attempt by an alpha male (●) or beta male (○) was interrupted before a successful copulation occurred, depending on the distance to the other male when the copulation display began. Data from 9 alpha-beta pairs. $n = 169$ alpha male attempts and 97 beta male attempts. From Davies 1985.

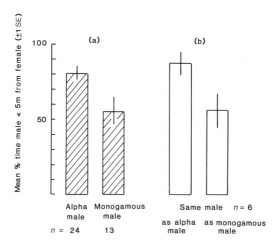

Fig. 6.3 Intensity of mate guarding. (a) Hatched columns; comparison between alpha males in polyandry and monogamous males (Mann-Whitney U-test, $P = 0.019$, 2-tailed). (b) Open columns, matched comparisons of same male involved in the two mating systems in different breeding attempts (Wilcoxon matched pairs test, $P = 0.05$).

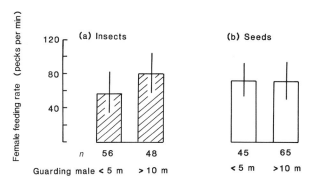

Fig. 6.4 Example from one female of the influence of male guarding distance on female feeding rate. Closer guarding significantly reduced female feeding rate when she was foraging on (a) small insects, hatched columns (*t*-test, $P < 0.001$, 2-tailed), but not on (b) seeds, open columns (NS). (n = number of measures of feeding rate, scored as pecks per minute; bars are 1SD.) Data from four other females all showed the same significant decrease for insect foraging ($P < 0.001$ in each case), and the one other data set for seed eating showed no effect.

feeding rate. However, this explanation is unlikely because there was no variation across mating systems in the time from the start of nest building to the laying of the first egg.[1]

Variation in guarding success by alpha males

Whether a beta male was successful in copulating with the female depended on how closely the alpha male was able to guard her (Fig. 6.5). In some cases, the alpha male kept within 5m of the female for almost all the time and beta males gained no access. In other cases, however, alpha males often lost the female and beta males were successful in mating. What causes the wide variation in the ability of alpha males to guard the female shown in Fig. 6.5? Four factors seemed to be important.

(a) *Territory characteristics.* As shown in Chapter 4, female territories varied markedly in size and vegetation cover. Beta males were more likely to succeed in mating on large female territories and on territories containing large patches of dense vegetation, such as dense shrubs and undergrowth or giant hogweed *Heracleum mantegazzianum* (Fig. 6.6). Presumably, the larger the territory and the denser the vegetation, the harder it was for the alpha male to maintain exclusive access, and the easier it was for the beta male and female to hide unnoticed.

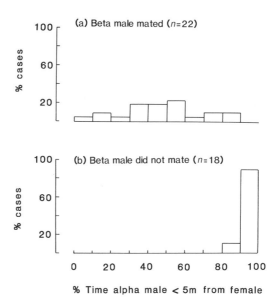

Fig. 6.5 Beta males were more likely to succeed in mating with the female when alpha males failed to guard her closely. Data from 40 females in polyandry and polygynandry. Difference between (a) and (b) highly significant (Mann-Whitney U-test, $P < 0.001$, 2-tailed).

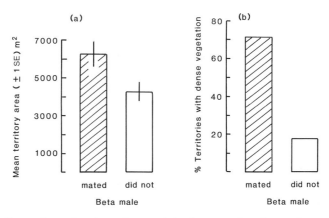

Fig. 6.6 Comparison of territory characteristics in cases where beta males succeeded in mating (hatched columns, $n = 14$) and those where they failed to mate (open columns, $n = 17$). Territories where beta males succeeded were significantly larger (a), Mann-Whitney U-test, $P < 0.05$, 2-tailed, and more likely to contain dense vegetation (b),
$$\chi^2_1 = 7.04, P < 0.01.$$

(b) *Beta male persistence.* Some alpha and beta males seemed more equally matched in competitive ability and although the alpha male could displace the beta male, he would not retreat more than a few metres and continually probed, provoking endless chases. Sometimes, these well-matched males seemed to call a temporary truce and broke off from chasing to feed side by side when I offered them some soft-bill food as respite. However, these quiet periods never lasted long and the two would soon be off again, playing 'cat and mouse' around the female. Some of these beta males gained as much, or even more, mating access as the alpha males. In other cases the two were so intent on competition that neither gained much access and the female hid, feeding alone, as if to avoid the conflict. Such variation provided a valuable source of data for unravelling how patterns of access to the female influenced male parental effort (see Chapter 12).

In other cases the beta males hardly probed at all and seemed to defer to the alpha male's superiority after a day or two's confrontation at the start of the mating period. Some gained no access and spent less than 5% of their time in the vicinity of the alpha male and female, simply feeding quietly alone in another part of the territory.

(c) *Polygynandry: activities of other females.* In polygynandry, where an alpha and beta male defended two or more females, the ability of the alpha male to monopolise matings depended on the laying synchrony of the females. Because each female had an exclusive territory, it was impossible for an alpha male to guard more than one at a time. Hence, whenever he was with one female, the beta male enjoyed free access to the others. I shall discuss how males allocate mating effort between females in Chapter 11.

(d) *Female behaviour.* Females not only relied on the alpha male to maintain contact during guarding, they also made life difficult for an alpha male by actively attempting to escape his close attentions and by encouraging the beta male to mate! Females often suddenly flew off low through the vegetation, twisting and turning as if to lose the alpha male. If he was busy feeding, he sometimes failed to see her go and meanwhile she approached the beta male and sought matings from him. If the alpha male did follow the female on her escape flight, she sometimes suddenly doubled back behind a bush and flew in the opposite direction in an apparent attempt to throw him off. Even when an alpha male was guarding closely, a female sometimes hopped quietly round the other side of a bush or rock and quickly solicited a mating from the beta male before the alpha male appeared again from behind the vegetation. On several occasions I saw females hiding away with the beta male under a hedge or bush. When the alpha male came by searching for them, they crouched down and remained motionless until he had passed by. There was no doubt at all that females were intent on preventing the alpha male from monopolising mating access.

These exploits were enormously entertaining to watch and it was difficult to think of the birds other than as scheming tacticians, each trying to outwit the others in the game. As I shall show in the following chapters, the outcomes of these so-called games had a vital influence on the participants' reproductive success. A male helped to feed a female's young, only if he had copulated with her earlier on during her mating period. The active attempts by the female to get the beta male to mate with her, as well as the alpha male, therefore represent an attempt to gain increased male help with chick rearing. The idea that female promiscuity may be a behavioural tactic to gain increased male care has been suggested for other species too (Dow 1977; Stacey 1979; Craig 1980; Briskie in press). For the males' part, I shall show how their conflicts also make good sense because of the costs of sharing paternity. The rules of the game, in the sense of why selection has designed the players in these ways, will thus become clear in later chapters. Understanding these rules has made the play all the more entertaining and as much pleasure can be gained from watching three dunnocks at mating time as from a good football match!

6.5 Duration of the mate guarding period

Where only one of the females in polygynandry is in her mating period (from end of nest completion to incubation), and where none of the other females have young, both alpha and beta males spent all their time competing for matings in her territory (Chapter 11). I have therefore included these cases together with polyandry in the following analysis, as there were no differences in behaviour between them. Figure 6.7 shows that there was no variation in the intensity of mate guarding during the mating period in either these trios or in monogamous pairs. In both, guarding continued right up to the last day before incubation began. For trios where alpha and beta males both gained some 'exclusive access' to the female (defined as the presence of only one male < 10m from her), there was no change through time in their share of access. Thus, alpha males were not more likely to monopolise females at any particular time during the mating period (Fig. 6.7b).

Eggs are fertilised about 24 hours before they are laid (Lake 1975). Thus, the last egg (number 4 in Fig. 6.7) will be fertilised at dawn, soon after the laying of the third egg. This means that guarding and copulations throughout the last day before incubation (day 3 in Fig. 6.7) are worthless. Why, then, do males continue to guard females on the last day even though there are no more eggs to fertilise? One explanation is that males simply do not know that the egg to be laid next morning will be the last one. Females normally began incubation on the morning the last egg was laid, but occasionally they delayed until the day after. Table 6.1 shows that males were clearly less interested in a female on the last day if she was no longer carrying an egg. There was less guarding by both

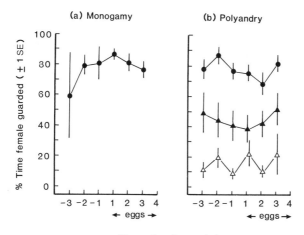

Fig. 6.7 Mate guarding relative to stage of mating period for cases where females laid a clutch of 4 eggs (days 1–4) and began incubation on the morning the last egg was laid (day 4). (a) Percent time monogamous males spent < 10m from female. (b) Percent time one or both of the alpha and beta males were < 10m (●) and percent time that alpha males (▲) and beta males (△) had exclusive access (only male < 10m of female) in cases where both males gained some access. Data from 8 monogamous males, 17 alpha males and 17 beta males. There was no significant variation with time in any of these relationships (Kruskal-Wallis 1-way ANOVAs, $P > 0.30$ in all cases). Data derived from Hatchwell and Davies 1992a.

monogamous and alpha males and less chasing between alpha and beta males in trios.

This suggests that males can detect when females are no longer carrying an egg but, when females are carrying an egg, cannot assess whether it will be the last one. Hence, males simply guard the female until she has no more developing eggs inside her. Thus it is unlikely that females could deceive males into providing parental care by delaying the onset of incubation, to increase the chance of them copulating at a time which would in fact be worthless. If such deception was possible, it would be of particular advantage to polyandrous and polygynandrous females, because alpha males sometimes prevented beta males from gaining access. However, there was no evidence that these females were more likely to delay the start of incubation (14.1% did so, $n = 92$) than monogamous females (7.1% did so, $n = 28$).[2] An experiment by Jones (1986) with sand martins suggests that males may detect egg carrying females by the way the extra weight influences their movements; when females were given small weights, males found them more attractive for mating.

Female dunnocks have sperm stores (see later) and, as in other passerines, it is likely that sperm can remain viable in their reproductive tract for a week or

Table 6.1. Comparison of mate guarding behaviour on the last day before incubation begins for cases where incubation began on (a) the day the last egg was laid and (b) the day after the last egg was laid. In (a), the female was carrying a developing egg on the last day before incubation. In (b), she was not. Figures in the table are means ± 1 SE (*n*). Guarding refers to distances of < 10m (from Hatchwell and Davies 1992*a*).

Behaviour of males	(a) Incubation began day last egg laid	(b) Incubation began day after last egg laid	P
Monogamous males % time guards female	82.3 ± 5.0 (11)	36.6 ± 4.7 (2)	< 0.05
Trios (polyandry and polygynandry) % time alpha male guards female	72.4 ± 7.1 (24)	34.0 ± 10.7 (10)	< 0.005
% time alpha, beta or both < 10m from female	83.4 ± 5.2 (24)	50.3 ± 12.6 (10)	< 0.02
% minutes with alpha versus beta male chases	16.2 ± 5.3 (24)	1.4 ± 0.7 (10)	< 0.02

Significance levels refer to 2-tailed Mann-Whitney *U*-tests.

more (T.R. Birkhead 1988). Thus it clearly pays males to compete for matings as soon as females begin soliciting, even though it may be a week or more before the first egg is laid. This pre-laying period was usually from 2 to 9 days, but early in the season the onset of laying could be delayed by cold weather for up to 3 weeks after nest completion. Males nevertheless continued guarding and mating throughout such long spells. However, the frequency of chases between alpha and beta male increased sharply the day before the start of egg laying (Fig. 6.8). This supports the idea that males can detect when females have a developing egg and compete most intensely during the period of peak fertility.

I do not have quantitative data on whether the intensity of guarding was influenced by time of day, but males certainly continued close guarding throughout all the daylight hours, from dawn to dusk. Guarding males roosted close to their females and were seen following them and copulating at first light the next morning. Females laid within the first 3 hours after dawn, and the males followed them to the nest and waited on a high perch nearby to overlook the nest during the 20 minutes or so the female was laying. When the female left the nest, the male flew down to follow close behind and attempted to maintain contact throughout the day.

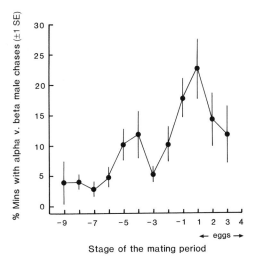

Fig. 6.8 Mean proportion of total observation time that alpha and beta males spent in chases relative to stage of mating period, for cases where females laid a clutch of 4 eggs (days 1–4) and began incubation the morning the last egg was laid. Data from 17 alpha-beta male pairs. From Hatchwell and Davies 1992a.

6.6 The value of mate guarding: removal experiments

In 1989 and 1990, Ben Hatchwell and I tested the importance of mate guarding for maintaining exclusive access to females by temporary removals of males for periods of 2–9 days during the mating period. Males were kept in aviaries and then returned to the Garden, where they all settled back on their territories. Figure 6.9 shows that when alpha males were removed, beta males gained increased access to the female, up to the same level previously enjoyed by the alpha male. Likewise, when monogamous males were removed, neighbouring males trespassed and gained increased access to females.

The beta males took over very quickly; in most cases they were already escorting the female while we were taking the alpha male out of the mist net! They then defended the female successfully against neighbours, who gained no more access than when the alpha male was in charge (Fig. 6.9). In monogamy, by contrast, there was often a delay in take-over of the female, particularly if the neighbouring males were busy guarding their own females (Hatchwell and Davies 1992a). During the mate guarding period, monogamous pairs often hid away in the vegetation and were difficult to observe. After removal of the male, the female also foraged quietly and so it may have taken neighbours some time to realise that there was an unguarded female available. Males were certainly

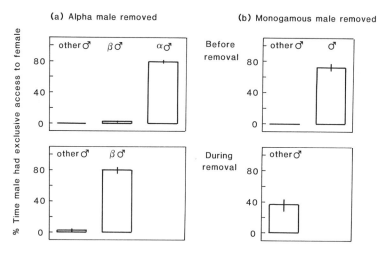

Fig. 6.9 (a) When alpha males were removed ($n = 26$) beta male access to the female increased. Comparing beta male access before (top) versus after removal (bottom), $P < 0.001$. Neighbours' access did not change (NS). (b) When monogamous males were removed ($n = 17$), neighbouring males gained increased access. Comparing access before (top) with after removal (bottom), $P < 0.001$. (Bars are 1SE and significance tests refer to Wilcoxon matched pairs tests.) From Hatchwell and Davies 1992*b*.

aware of the mating activities on neighbouring territories. They often perched high on tree tops, apparently watching what was going on next door, and occasionally made surreptitious visits to neighbouring nests to inspect their contents.

These removals, like those with other species (Björklund and Westman 1983; Møller 1987), showed clearly that mate guarding was necessary to prevent other males from gaining matings with a female.

6.7 Female responses to individual male songs

Our observations of females during the mating period suggested that they could recognise males by their song and used male song to locate the individuals they wanted to copulate with. When a female was guarded by an alpha male, she often flew towards the beta male when he sang, but when she was accompanied by the beta male she did not approach the singing alpha male (Table 6.2). This is the behaviour we might expect if females were intent on gaining copulations from both alpha and beta males under conditions where alpha males were attempting to monopolise all the matings. When females were alone they approached the song of either of the resident males, alpha or beta, and monogamous females likewise approached their own male when he sang. By

Table 6.2. Responses of females to singing males during the mating period. Each female was observed for 1–15 hours from nest completion to incubation. The numbers of females for each case vary because not all situations occurred during the observations of each female (from Wiley *et al.* 1991).

Female's social context	Singing male	Proportion of females observed approaching	(n)
Polyandry and polygynandry			
With alpha male	Beta male	0.26*	(38)
	Neighbour	0	(21)
With beta male	Alpha male	0	(29)
	Neighbour	0	(6)
Alone	Alpha male	0.25†	(51)
	Beta male	0.25	(40)
	Neighbour	0	(15)
Monogamy			
With male	Neighbour	0	(9)
Alone	Mate	0.59**	(17)
	Neighbour	0	(4)

* More likely to respond to beta male than neighbour ($P < 0.01$, Fisher exact test).
† No difference between response to alpha and beta ($\chi^2_1 = 0.04$), but female responds to both more than to neighbour ($P < 0.01$ in each case). ** More likely to respond when alone to mate than to neighbour ($P = 0.055$, Fisher exact test).

contrast, females never approached neighbouring males, even when they trespassed and sang on the female's territory (Table 6.2). This makes good sense too, because neighbours do not provide parental care even when they succeed in copulating (Chapter 7).

Haven Wiley recorded male songs during the summer of 1989 and found that when males were singing alone in their territories they used repertoires of up to seven different song patterns, each consisting of a stereotyped sequence of unique notes, which altogether lasted 4 seconds or more. A male usually sang the same pattern repeatedly before switching to another pattern. Successive songs of the same pattern, however, differed in how much of the complete sequence was sung so most songs lasted only 1.5 to 2.5 seconds. Wiley found that males who shared a territory, and males on neighbouring territories, often shared portions of one or more of their song patterns, but in these cases males sang individually distinct variants of the shared pattern (Fig. 6.10). These findings are similar to those of D.W. Snow and B.K. Snow (1983).

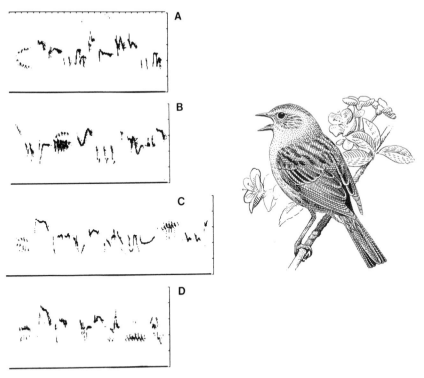

Fig. 6.10 Spectrograms of male dunnock song (vertical divisions 2kHz; hortizontal division 62.5ms). A, one of male YNL's song patterns; B and C, songs of two males on neighbouring territories; D, another of YNL's song patterns to show resemblances to a pattern of song C of his neighbour. In one playback experiment during YNL's temporary removal, his female approached the speaker during playback of song A but not during playback of song C. From Wiley *et al.* 1991.

We used playbacks during the removal experiments described in the previous section to test whether females really could use song alone to discriminate among males (Wiley *et al.* 1991). During the period the alpha male or monogamous male was removed, we presented tape-recorded songs to females. In one trial we broadcast the song of their missing mate and in the other, the song of a neighbouring male. The trials were in random order and separated by 15 minutes. Each playback consisted of two repetitions of a sequence of four consecutive songs recorded from the field. The presentations were done on either the first or second day after the female's mate was removed. Females were scored as responding to the playback if they either approached the speaker or gave a 'triplet call', a rapid high pitched trill 'ti-ti-ti', often given when females are separated from their mate.

Ten females were tested during their mating period. Six were monogamous; four responded to the song of their missing mate and not to the neighbour's song; one responded only to the neighbour's song and one showed no response to either. Four were polyandrous females; two responded to the song of their missing alpha male and not to the neighbour and two failed to respond to either. Of the seven responses, two were simply calls and five were approaches to within 3m of the speaker. All five of these strong responses were to playback of the missing mate's song.[3]

These experiments confirm that females can recognise their mates by their songs alone. Furthermore, the responses seemed to be behaviour specifically concerned with gaining copulations. Whereas 6 of the 10 females tested during their mating period responded to their missing mate's song, none of the 6 females tested outside their mating period (during incubation or before nest building) did so.[4]

6.8 Copulations

Copulation rate

In many birds, particularly passerines, copulations are rarely seen. They occur at a low rate and often out of view. For example, David Snow (1988) saw only five during his detailed four year study of blackbirds *Turdus merula* and Glen Woolfenden and John Fitzpatrick (1984) saw about twenty in their ten year study of the Florida scrub jay. Dunnocks, by contrast, copulate at an extraordinarily high rate and one can observe as many copulations on a dunnock territory in a single morning as in years of intense study of most other species. I shall show how this is not only a reflection of the intense competition for paternity, particularly among alpha and beta males, but also results from the active part played by the female in soliciting matings at a high rate.

There was no variation during the mating period in the copulation rate, for either monogamous pairs or alpha and beta males in polyandry or polygynandry.[5] Overall rates were obviously affected by the frequencies of interruptions and chases among alpha and beta males. To allow comparison between mating systems, the rates in Table 6.3 refer to rates for periods of peace, where a male and female associated alone together, with no other male within 10m. These data show that the copulation rate is not simply related to the number of males on the territory, but rather to the success of competitors. There was no difference between monogamous pairs and trios where beta males failed to gain access to the female, but a marked increase in rate where both males gained access. Therefore, the presence of a beta male on the territory did not in itself cause alpha males to increase their copulation rate, only the presence of a beta male who was succeeding in mating with the female.

Table 6.3. Copulation rates of males and females in the different mating systems. Rates for females are for periods of uninterrupted association with one resident male. Rates for males are likewise for periods of uninterrupted access, when the male was the only male < 10m from the female (data derived from Hatchwell and Davies 1992*a*).

Mating system	Sex	Copulations per hour (mean ± 1 SE)
(1) *Monogamy* (*n* = 14)	Female/male	0.47 ± 0.17
(2) *Polyandry and polygynandry* (a) Only alpha male obtained exclusive access (*n* = 17)	Female/male	0.87 ± 0.16
(b) Both alpha and beta males obtained exclusive access (*n* = 32)	Female Alpha male Beta male	2.02 ± 0.27 2.45 ± 0.47 2.40 ± 0.31

No significant difference between (1) and (2a), but rates in (2b) significantly higher for females than in (1), $P < 0.001$, and than in (2a), $P < 0.01$. Rates for alpha males significantly higher in (2b) than in (2a), $P < 0.01$ (Mann-Whitney U-tests, 2-tailed). No significant difference between alpha and beta males in (2b) (Wilcoxon matched pairs test).

A more detailed analysis of the influence of mating competition is shown in Fig. 6.11, which plots copulation rates against the proportion of 'exclusive access time' enjoyed by the beta male (i.e., periods when no other male was < 10m, so the male could associate with the female in peace and gain successful copulations). For both alpha and beta males, the rate of copulation increased with the proportion of access gained by the competitor. For females, the rate reached a peak at around the 50:50 share of access between the two males.

There is a simple explanation for how this increase in rate with competition comes about. A male is most likely to copulate when he first joins a female, at the start of a period of exclusive access (Fig. 6.12). The rate then drops, with alpha and beta males showing the same decline. With increased competition between alpha and beta males, the female swaps partners more frequently, as first one male gains access and then the other does so. The overall copulation rate increases with the rate of partner swapping, because each male copulates at a high rate at the start of each bout of access.

Active solicitations by females

The copulation rate is not only influenced by competition among the males. It would be impossible for a male to force a mating with a female, as occurs in

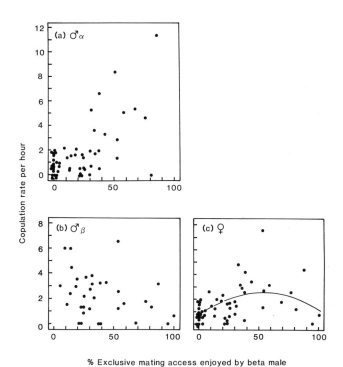

Fig. 6.11 Copulation rates per hour of exclusive access (only male < 10m) for (a) alpha males, (b) beta males, and (c) females, in all cases of polyandry and polygynandry, relative to the proportion of exclusive access gained by beta males. Copulation rates of alpha males increased ($r_s = 0.418$, $n = 56$, $P < 0.01$) and those of beta males tended to decrease ($r_s = -0.291$, $n = 34$, $P < 0.10$) as the share of beta male access increased. Female copulation rate peaked at about 50:50 alpha:beta access ($F = 9.86$, 1,55 df, $P < 0.01$). From Hatchwell and Davies (1992a).

some ducks and seabirds (Cheng *et al.* 1983; McKinney *et al.* 1983), because female dunnocks must first present their cloaca in a special posture to facilitate copulation (see below). The high rates occur, therefore, with full acquiescence by the female. In fact her behaviour suggests that she would accept an even higher rate than the males are prepared to provide. Females usually solicited whenever a resident male approached and even during quiet periods, when they associated with one male, they solicited every few minutes. During periods of uninterrupted mating access with one of their males, polyandrous females solicited on average every 19 minutes, while the interval between copulations was 28 minutes—significantly longer.[6] Males thus ignored many of the solicitations, especially those occurring soon after a copulation (Hatchwell and Davies

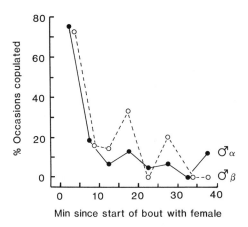

Fig. 6.12 Proportion of occasions alpha (•) and beta males (○) copulated with a female in successive 5min periods from the start of a bout of uninterrupted access, following a period of separation from the female. Data from 39 different males and 29 different females. Data derived from Hatchwell and Davies 1992*a*.

1992*a*). The females' attempts to escape alpha male guarding and their approaches to beta males to solicit matings (described above) also clearly show the active part they played in determining copulation rate.

Females may also have played an active role in avoiding the attentions of neighbours. During our removals of monogamous males we noticed that the females seemed reluctant to copulate with the neighbouring males when they first came onto the territory. Some flew off into dense vegetation, as if to escape the newcomer's advances, and others displayed aggressively to the males, fluffing up their body feathers and making darting runs towards them as if to drive them away. Only after several hours persistance, did the new males succeed in copulating. These observations, together with the song playbacks, show that females have no interest in gaining matings from neighbours, only from the resident males on the territory. This supports the view that females offer matings, not only to fertilise their clutch, but also to get males to provide parental care, with only permanent residents offering this benefit (Chapter 7). I never saw females seeking extra-bond matings in the way described for some other species (S.M. Smith 1988; Møller 1988*a*), where it has been suggested that females may go shopping for good genes.

6.9 An extraordinary pre-copulation display: cloaca pecking

The strange pre-copulation display of the dunnock was first described in 1902 by Edmund Selous and recorded as follows in his book *Evolution of Habit in*

Birds (1933, pp. 107−9). 'I made a curious and very intimate observation on the nuptial habits . . . A pair of these little birds came out from the shrubbery . . . and, after hopping about a little, in a state of great sexual excitement, the hen elevated her rump and stood still, when the male, hopping up, made little excited and very wanton-looking pecks in this region, that is to say into the actual orifice. There was actually no mistaking the nature and significance of the actions, rather lecherousness, as it seemed to me, being revealed . . . This is a very remarkable thing to have seen, I think . . . But I do not understand it.' Later, the display was celebrated in a series of notes to the journal *British Birds*, all entitled 'Extraordinary display by a pair of hedge sparrows' (Delamain 1929; Rollin 1929; Gardam 1929; Clark 1930; Orton 1930; Meiklejohn 1930; Hartley 1930). The editor terminated this long correspondence when it was realised that, although certainly extraordinary, the display was part of the species normal behaviour and precedes every copulation (see also Harrison and Binfield 1967).

The male hops about behind the female, and she quivers her wings and raises her tail so as to expose her cloaca. The male then pecks at her cloaca on average 30 times (range 0−90), for a period of 57 seconds (range 8−160sec),[7] before he finally copulates. During this long display there is an increased chance that another male will interrupt before copulation itself occurs. On territories where there was a single male guarding the female, only 2% of copulation attempts were interrupted during the pecking display (by neighbours) but where alpha and beta males were competing for matings on average 22% of the displays were interrupted by the competing male.[8] On one territory with a particularly persistant beta male, 36 of the 46 copulation attempts I observed by the alpha male were interrupted during the cloaca pecking, with the beta male flying in close to the displaying pair and the alpha then leaving the female to chase him off. This makes one wonder why on earth the male does not, like most birds, simply copulate quickly straight away to ensure sperm transfer. The pecking obviously must be very important to offset these interruption costs.

During the pecking, the female's cloaca becomes pink and distended and makes strong pumping movements. From time to time, she dips her abdomen down suddenly and occasionally ejects faeces. On other occasions, however, she ejects a small droplet of fluid (seen also by Sanderson 1968). There is little doubt that this is what the male has been waiting for, because he looks at the droplet briefly and, as soon as it is produced, he copulates. Copulation itself is brief; the male jumps at the female, cloacal contact lasting for only a fraction of a second. This differs markedly from most passerines, where the male perches on the female's back for several seconds and makes repeated cloacal contact.

I can still remember my intense excitement when I first noticed a droplet being ejected and suddenly realised that the function of the display might be concerned with a battle between males for paternity and the droplets might be sperm from previous matings. I immediately got on my hands and knees and began to search for the droplets. Even when the display had been watched from a close distance

TOP, prior to copulation the female raises and quivers her tail from side to side, while the male pecks her exposed cloaca. MIDDLE, during the pecking, the female dips her abdomen and sometimes ejects a small droplet of fluid containing sperm from previous matings. The male cocks his head briefly to look and then, BOTTOM, he copulates, cloacal contact lasting a fraction of a second.

and, through binoculars, I had been certain of locating the exact spot where the droplet landed, it was a difficult job to find it, and I managed to discover only three. In each case the droplet had landed by chance on a stone or small pieces of bark, which held it intact like a spoon. I rushed back to the laboratory and was thrilled to see, through a microscope, that the droplets indeed contained masses of sperm (Fig. 6.13).

It is impossible to tell from field observations exactly how often droplets are ejected, but my interpretation of the display was as follows (Davies 1983). Female birds have sperm stores at the utero-vaginal junction adjoining the cloaca. With frequent copulations from two males, a female's sperm stores might become full. The copulation rate of dunnocks is extraordinarily high, averaging in some cases twice per hour throughout a ten day mating period. If a female's sperm stores become full, then the only way a male could put sperm in would be first to stimulate the female to eject some to make room. Sperm in the store may be another male's, whereas sperm a male is about to put in is certainly his own. We could speculate still further that it may pay the female to eject sperm in front of a male to convince him that he would have a chance of paternity, and so encourage him to feed the young.

Although the data from the first two years of the study suggested that the pecking display was more intense where two males competed most vigorously for matings (Davies 1983), the effect was not strong and a subsequent analysis on a larger data set has shown no variation in display duration, or number of pecks, with mating system, overall copulation rate, or time in the mate guarding period (Hatchwell and Davies 1992*a*). A male's display performance was also not influenced by whether he himself or another male was the last to copulate. In contrast to mate guarding intensity and copulation rate, therefore, the display did not vary in relation to the degree of competition for matings.

6.10 Sperm competition and the reproductive organs of males and females

Tim Birkhead examined whether male and female dunnocks have any peculiarities in their reproductive organs which are related to their extraordinary sex life, with the extremely high copulation rate and bizarre mating display (Fig. 6.14). One bird of each sex was dissected, both killed by cats during the mating period (for details see T.R. Birkhead *et al.* 1991).

Male

In the breeding season male dunnocks have an unusually large cloacal protuberance (Fig. 6.15) and a small fleshy eversible phallus which, by manipulation of the cloaca, we could extrude 2–3mm. The phallus may aid sperm transfer during the remarkably brief contact between male and female

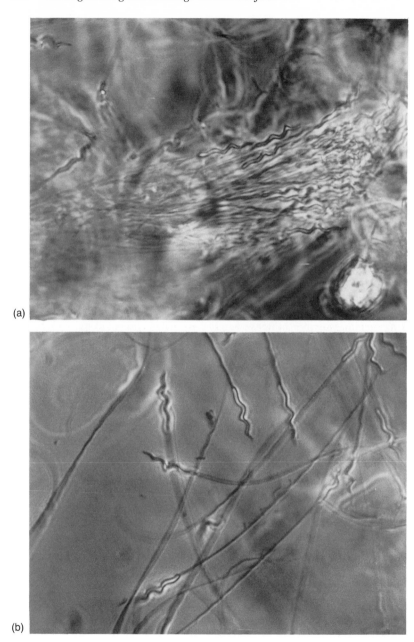

(a)

(b)

Fig. 6.13 (a) Part of a sperm mass ejected by a female during cloaca pecking by the male. The semen of passerine birds is a small glutinous droplet, with little accessory fluid, containing bundles of sperm, which have helical heads. (b) Some individual sperm on the edge of the mass. From Davies 1983.

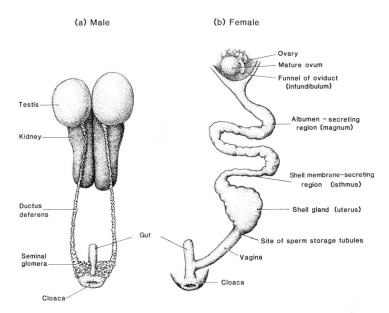

Fig. 6.14 Schematic diagrams of (a) male and (b) female reproductive organs showing the seminal glomera tubules in the cloacal protuberance of the male and the site of sperm storage tubules in the female, at the utero-vaginal junction.

during the act of copulation itself. Dissection showed that the cloacal protuberance comprised mainly the seminal glomera, extensively coiled tubules of the ductus deferens which transports sperm from the testes. The tubules were packed with sperm, estimated to number $1,060 \times 10^6$. The only comparable data are for two monogamous estrildine finches, the Bengalese finch *Lonchura striata* and the zebra finch *Poephila guttata*, where the number of sperm in the seminal glomera was a thousand times less (T.R. Birkhead in prep). Although these two species are slightly smaller (16–17g) than the dunnock (20g) this difference is still dramatic. The weight of the two testes was 0.688g, 3.4% of the male's body weight, which is some 64% greater than that expected for a bird the size of the dunnock (Møller 1988*b*).

Thus the male dunnock is adapted morphologically for high levels of sperm production. The number of sperm in each ejaculate is not known, but based on the relationship with testes weight in other birds it may be around one million. Bengalese finches which have copulated three times in three hours seriously deplete their sperm stores (T.R. Birkhead in prep). Dunnocks, with reserves a thousand times as large, are obviously capable of frequent copulation. With sperm produced continuously in the testes to keep the seminal glomera topped up, it seems likely that they can transfer sperm at the observed copulation rate

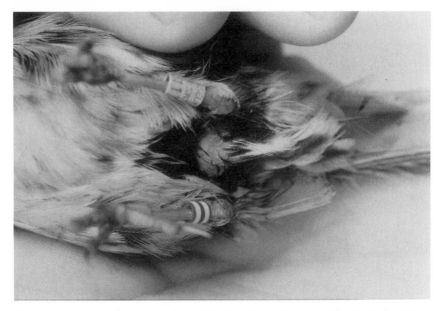

Fig. 6.15 Photograph showing the male's cloacal protuberance. The bird's head is to the left.

of 7 times a day for monogamous males and 14 to 28 times a day for alpha and beta males in polyandry and polygynandry (calculated from Table 6.3).

Female

Female dunnocks have swollen cloacas during the mating season, but no protuberance similar to that in males. As in the male, the cloaca is surrounded by *c.* 15 small feathers. During the pre-copulatory display these are folded back to expose the cloaca. Dissection showed that the utero-vaginal junction contained about 1,400 sperm storage tubules, narrow blind ending tubes each about $12\mu m$ in internal diameter and $370\mu m$ in length. Ninety per cent of the tubules contained sperm, with *c.* 500 in those most densely packed (Fig. 6.16). The morphology, number and size of these tubules is similar to that described for other passerines (Shugart 1988; T.R. Birkhead and Møller 1991), so despite the high copulation rate the female dunnock does not seem designed to store an unusually large quantity of sperm.

Given the limited ability of females to store sperm, and the huge capacity of males to inseminate females, it makes good sense for males to stimulate females to eject sperm before they copulate. However, the origin of the ejected sperm may not be the sperm storage tubules themselves. We did not see any contractile elements which might facilitate the release of sperm from the tubules, nor have these been found in other species (T.R. Birkhead and Møller 1991). In poultry,

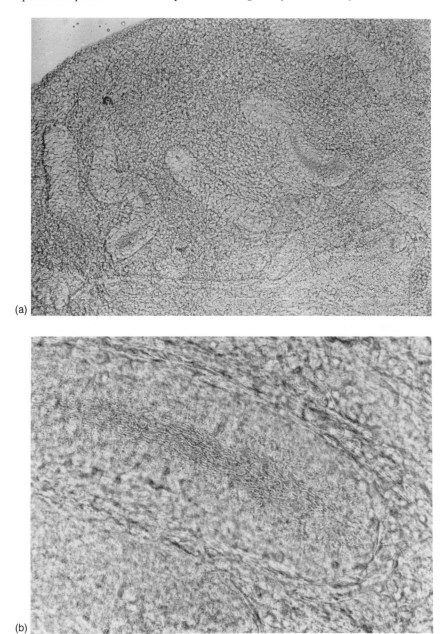

Fig. 6.16 Sperm storage tubules in the female's utero-vaginal junction. (a) Four tubules. The first and third from the right are packed with sperm. The others are empty. (b) Closer view of a tubule containing sperm. Tubules are on average 370μm in length. From T.R. Birkhead *et al.* 1991.

sperm probably take several hours to move from the vagina into the sperm storage tubules (Bobr *et al.* 1964; Verma and Cherms 1965) and it seems likely that in dunnocks, too, there will be a pool of sperm in the female's vagina for much of the time, especially given her high copulation rate. The most likely function of the pre-copulation display is for males to stimulate females to eject sperm from the cloaca and vagina to make way for their insemination, and so give their own sperm a better chance of reaching the sperm storage tubules. It would also clearly pay a male to engage in this display and copulate as soon as he encountered a female after a period of absence (as in Fig. 6.12), in order to prevent sperm that may have come from another male reaching the sperm storage sites.

Comparison with the alpine accentor

Alpine accentors have a polygynandrous mating system and females also copulate at a high rate with several males. The males have large testes and cloacal protuberances, as in dunnocks, and also seem adapted for high rates of sperm production (Fatio 1864; Aichhorn 1969; M. Nakamura 1990). In this species the female likewise plays an active role in soliciting copulations from several males.

The pre-copulatory display shares some similarities with that of the dunnock, but there are interesting differences (M. Nakamura 1990). The female presents her exposed cloaca to the male for up to 30 seconds prior to copulation, and the cloaca also makes strong pumping movements. However, the male does not peck the female's cloaca, but simply stands behind her in an upright posture and watches. As in dunnocks, copulation involves extraordinarily brief cloacal contact, with the male leaping at the female's abdomen and the female then also leaping away immediately after contact. Analysis of film of captive birds showed that the male clings to the female for only about 0.15 seconds, one foot on her back and the other on her belly, and beating his wings for balance. It seems likely that when the male presses his cloaca against the female, the sperm will be expelled immediately from the seminal glomera (Aichhorn 1969). Unlike dunnocks, the female then immediately presents her cloaca to the male again and he makes cloacal contact up to five times in quick succession. It remains to be shown whether, like dunnocks, females ever eject sperm.

6.11 Summary

Males guarded their females closely throughout the mating period, from the time the female completed the nest right up to the laying of the last egg. Monogamous males chased off neighbours and alpha males guarded females even closer, in an attempt to prevent beta males from mating. Beta males competed intensely to gain access to the female and were more likely to succeed when territories were large or had dense vegetation, because alpha males were then more likely to lose contact with the female.

Females also made it difficult for alpha males to monopolise them by active attempts to escape their close guarding and by encouraging the beta male to mate. Females often approached beta males when they sang, and playbacks showed that females could distinguish the songs of their resident males from those of their neighbours, approaching only the former. Approaches to playback occurred only during the female's fertile period and seem specifically concerned with seeking copulations.

The intensity of mate guarding and copulation rate did not vary through the mating period but males clearly detected when females were most fertile because male chasing increased the day before egg laying began and guarding was less intense when females delayed the start of incubation for one day after clutch completion.

Temporary removals of monogamous males resulted in greater mating access by neighbours and removals of alpha males led to increased beta male access, showing clearly that mate guarding was necessary to keep competitors at bay.

A male increased his copulation rate as the success of his competitors increased. Males were most likely to copulate when they first encountered the female after a period of absence. Males are adapted for high copulation rates, with unusually large testes and a cloacal protuberance packed with seminal glomera containing a massive amount of sperm. Females have sperm storage tubules at the utero-vaginal junction but these are similar in number to other species so females are not designed to store unusually large quantities of sperm.

There was an extraordinary and prolonged pre-copulation display, where the male pecked the female's cloaca and stimulated her to eject sperm from previous matings. The ejected sperm is probably from the cloaca and vagina. It may pay males to delay copulation until they have ensured that there is a clear passage for their own sperm to reach the female's sperm storage tubules.

STATISTICAL ANALYSIS

1. Mean number of days from the start of nest building to the laying of the first egg was 7.54 (n = 11) for monogamy, 9.17 (n = 12) for polyandry and 8.89 (n = 9) for polygynandry (Kruskal-Wallis 1-way ANOVA; NS).

2. G = 1.07, 1df, NS.

3. Binomial probability, 2-tailed, P = 0.06.

4. Fisher exact test, $P < 0.05$.

5. Data from 25 monogamous males, 63 alpha males and 37 beta males who obtained some mating access. Kruskal-Wallis 1-way ANOVAs showed no significant variation over the period 4 days before the first egg through to clutch completion and incubation (for details, see Hatchwell and Davies 1992a).

6. Interval between successive solicitations = 19.1min (SE 1.84, n = 63). Interval between successive copulations = 28.2min (SE 2.54, n = 40); t_{102} = 2.901, P < 0.01, 2-tailed. Data from periods of uninterrupted access by a male.

7. Data from 151 copulation displays involving 44 different males and 34 different females (from Hatchwell and Davies 1992*a*).

8. Of 14 cases of polyandry and polygynandry studied during 1981−1983, no difference in frequency of interruptions of alpha versus beta male copulation attempts but average proportion interrupted (21.7%, SE = 9.1) significantly greater than for monogamous pairs (mean 2.3%, SE = 2.4, n = 11), P < 0.05, Mann-Whitney U-test, 2-tailed. Total of 162 displays observed.

7

Relating behaviour to maternity and paternity

7.1 Male mating success and parental care

After all the excitement of the mating period, life suddenly becomes peaceful on a dunnock territory during incubation. While the female incubates the clutch, the males feed quietly and an alpha and a beta male are sometimes seen side by side with no signs of the serious squabbles which characterised the days when the female was soliciting copulations. It is as if a whistle had been blown to signal the end of a game.

When the young hatch, activity increases once more. On territories with monogamous pairs both male and female feed the young. On territories with alpha and beta males, however, there are markedly different patterns of chick feeding. In some cases both males can be seen hard at work helping the female to provision the young, often foraging side by side as they collect billfuls of small insects. At these nests, therefore, the young get fed by three adults. In other cases, however, only the alpha male helps the female and the beta male shows no interest in the brood at all. One of the most important discoveries in the first years of the study was that this marked difference in parental care by the beta males was related to their success in gaining matings with the female earlier on during the copulation period.

Table 7.1 shows how a male's success in gaining mating access with a female influenced his decision to feed her young. The data are from 19 cases of monogamy, 21 cases of polyandry and 48 cases of polygynandry where females were observed during the mating period to score which males gained access, and where the chicks hatched successfully so that parental care could also be measured. During the mating period, females were observed for 4.02 hours each, on average. This seemed sufficient to get a reasonable measure of the relative mating access of two males because in 85% of the cases where beta males were seen to gain some access to the female, they did so within the first hour of observation. Mating access was scored as time when a male was the only

Table 7.1. Influence of a male's success in gaining mating access to a female on his probability of feeding her young.

Mating system	No. cases male fed young	
	Gained mating access to female	Did not gain mating access
Monogamy		
Resident male	19/19	–
Intruding neighbour	0/3	0/16
Polyandry		
Alpha male*	18/18	–
Beta male	11/14	0/7
Intruding neighbour	0/5	0/16
Polygynandry		
Alpha male	30/44	0/4
Beta male	22/35	1/13
Intruding neighbour	0/1	0/47

* Three alpha males in polyandry killed by predators during incubation so their chick feeding could not be scored—all gained mating access. Beta males are more likely to feed the young if they gained mating access with the female in both polyandry ($\chi^2_1 = 8.614$, $P < 0.01$) and polygynandry ($\chi^2_1 = 9.453$, $P < 0.01$).

male less than 10m from the female, in other words, time when he could potentially mate with her, uninterrupted by other males (Chapter 6). I have scored male success in terms of exclusive mating access rather than copulations simply because males often hid away with females in dense vegetation and so copulations could not always be seen. However, mating access time certainly gave a good measure of copulation success (see below).

Provisioning of young was recorded when they were 5–11 days old, when feeding rates were highest. Nests were watched for periods of 0.5–2 hours duration, for a total of 4.35 hours per nest, on average. Again, this seemed sufficient to get a good measure of male provisioning effort because in 90% of the cases where we recorded a brood as provisioned by two males and a female, all three adults were seen feeding the chicks within the first hour of observation.

In monogamy, not surprisingly, the resident male always gained mating access and Table 7.1 shows that he always helped to provision the young. Likewise in polyandry, alpha males always succeeded in gaining some mating access and always helped feed chicks. Beta males, however, only sometimes succeeded in gaining a share of the mating access and were more likely to help feed the young if they did so (Table 7.1). Where both alpha and beta male gained mating access, therefore, both usually helped to feed the brood. The few cases where a beta male gained mating access but did not feed the young were all cases where his share of the matings was very small (Chapter 12). There was no influence of brood size; of the 14 cases in Table 7.1 where the beta male gained mating access he helped to feed the young in 2/2 cases where there was just a single chick in the nest, 2/3 cases where there were two young, 3/4 cases where there were three, 3/4 cases where there were four and 1/1 case where there were five young in the nest.

In three cases of polyandry the beta male failed to gain mating access in the first breeding attempt of the season and did not feed the young. In the second brood, however, when the vegetation had grown and the female could more easily escape the alpha male's close guarding, the beta male succeeded in mating and then did help to feed the young. A beta male's chick feeding response was therefore specific to his mating access in that particular breeding attempt.

In polygynandry, where the alpha and beta male shared several females, their mating access with one female, and their parental care of her brood, were both influenced by the activities of the other females in the system (Chapter 11). For example, alpha males occasionally restricted their attentions to one of the females and left the beta male to mate with the other, and sometimes the males left a female to care for the brood alone because they were competing for matings with other females. Nevertheless, despite these complications, the same clear trends are evident, as in polyandry. A male was more likely to feed the young if he had previously succeeded in gaining matings with the female (Table 7.1).

These results immediately suggest an explanation for why females encourage beta males to mate. By getting the beta male to mate, as well as the alpha male, the female gains parental help from two males rather than one, and so increases her total amount of help (Chapter 8). Table 7.1 shows, by contrast, that neighbouring males never helped to feed a brood, even in the few cases that they succeeded in gaining some mating access. This explains why females do not encourage neighbouring males to copulate. Later in the book I shall consider whether females might suffer a cost from allowing neighbours to copulate, because of reduced care from resident males who detect that they have been cuckolded (Chapter 12).

7.2 Does male mating access reflect paternity?

One of the problems of working in a public place, like a Botanic Garden, is that people often come up and ask what you are doing. When I am crawling about looking for sperm droplets or dunnock faeces, a truthful reply entails a rather detailed explanation to be convincing and so I usually announce that I am 'weeding' in order to avoid a long interruption.

My most memorable encounter was with a nun. I was at the top of a ladder checking a brood of nestlings one day when a small voice called out, 'I hope you are not disturbing those birds'. Down below was a nun, dressed in black and holding her rosary beeds. I felt obliged, in this case, to reveal the truth, and told her of the dunnock study. 'Why such interest in dunnocks?' she asked. I elaborated further, and risked the description of two males competing for mating with a female. 'That's most interesting, and which male fathers the offspring?' asked the nun. I then admitted that she had put her finger on the key question and that I did not know the answer. 'Well surely that should be easy to discover', she replied, 'why don't you analyse protein polymorphisms?' The nun turned out to be a chemistry teacher at the local convent school and this encounter made me realise that I really did need to find out how matings related to paternity if I was to understand the links between mating and chick feeding.

Polymorphic proteins, usually extracted from small blood samples, have been used as genetic markers to assign paternity in several studies of mammals (Schwartz and Armitage 1980; Foltz and Hoogland 1981; Hanken and Sherman 1981). The method is based on the observation that some proteins occur in several forms (morphs) which differ with respect to one or more of their constituent amino acids. The various forms of a protein differ in charge and molecular weight, which influences how fast they move through a gel. They can be identified, therefore, by gel electrophoresis and it is possible to check whether the morphs are under genetic control by looking at their ratios in offspring from matings between parents with known morphs. The more proteins (loci) that can be studied, or the more morphs per protein (alleles per locus), the better the chances that potential fathers will be distinct and so the easier it is to assign paternity.

This method has also been used with success in birds (Gowaty and Karlin 1984; Westneat 1987; Wrege and Emlen 1987; Sherman and Morton 1988). However, there seems to be less variation in blood proteins than in mammals and in many studies it has proved difficult to assign paternity with precision. For example, resident and neighbouring males may often share protein types so the frequency of extra-pair paternity may be underestimated. Even when a resident male can be excluded as the father, because a chick has a protein type different from both him and the mother, it is often impossible to distinguish which is the father among several other males.

7.3 DNA fingerprinting

A more powerful method is to look at variation in the genetic material itself, rather than variation in the proteins which it encodes. In 1985, Alec Jeffreys discovered that variability in human DNA could be used to provide each individual with a unique 'genetic fingerprint' (Jeffreys *et al.* 1985). This has provided a technique for unambiguous assignment of parentage in humans to solve paternity disputes and also enables identification of rapists from tiny amounts of sperm. The method has since been developed for birds (Burke and Bruford 1987; Wetton *et al.* 1987) and is now fast becoming as much a part of a scientific ornithologist's essential equipment as are binoculars and colour-rings (T.R. Birkhead *et al.* 1990; Gibbs *et al.* 1990; Gyllensten *et al.* 1990; Rabenold *et al.* 1990; Westneat 1990).

Unlike mammals, birds have nucleated red blood cells so a tiny drop of blood provides sufficient DNA for the analysis. The DNA is extracted from the blood and then cut into pieces with a restriction enzyme. These fragments are then spread out, according to size, by gel electrophoresis with smaller fragments moving faster and so reaching the bottom of the gel first (Fig. 7.1). Jeffreys' discovery was that there are particular sequences of nucleotides (the building blocks of DNA) which can be visualised to reveal a unique genetic fingerprint for each individual. The sequences are repeated in variable numbers, from ten to many hundred times, and are distributed on many chromosomes. Each sequence can be detected with a radioactively labelled probe to produce bands on an autoradiograph of the gel. Remarkably, the core part of each sequence, to which the probe binds, is the same in humans and birds, so the probes developed by Jeffreys can also be used for fingerprinting birds. What varies between individuals is the distribution of the sizes of fragments containing a particular repeated sequence. There is so much variation that no two individuals in a population are alike, so everyone has a unique pattern of bands or fingerprint.

Terry Burke and Mike Bruford, from Leicester University, were among the first to pioneer the use of DNA fingerprinting for studies of animal populations and we were very lucky when they agreed to join the dunnock study, with Ben Hatchwell and myself providing the samples and behavioural observations to link up with the paternity analysis. Our collaboration began in 1988, when Ben and I collected a small drop of blood, from the brachial vein, from all the adults in the breeding population (caught by mist net) and all their nestlings, which were sampled when they were 6–7 days of age. The sampling, done under licence from the Home Office and Nature Conservancy Council, took just a few minutes and caused no obvious distress to the birds, and we had no evidence that it ever did them harm. The adults immediately resumed their normal behaviour and were sometimes copulating or defending their territories within minutes of release, while the nestlings began begging as soon as we put them back in their nests.

Our first job was to analyse a large family to see how the fingerprint bands were inherited. Figure 7.1 compares fingerprints of a male and female of a monogamous pair, together with those of the 11 offspring they raised in a year, from 3 broods. The figure shows that each individual has a unique pattern of 20–30 bands. A detailed analysis revealed that almost all the bands were inherited independently, with each offspring getting a random assortment, half

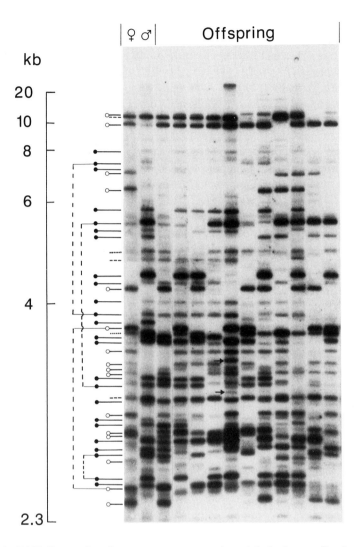

Fig. 7.1 DNA fingerprints from a monogamous pair and their eleven offspring, from three broods, in 1988. The spread of different-sized DNA fragments, shown as bands, provides a unique 'genetic fingerprint' for each individual. The scale at the side is in kilobases, larger fragments at the top and smaller ones at the bottom. Segregation analysis of the 16 maternal (○) and 24 paternal (●) fragments revealed a transmission frequency to offspring of 0.495, four apparently allelic pairs (joined by dashed lines) and no consistent cosegregation. All bands in all the offspring are shared with either the male or female, except for two arrowed bands in the fifth offspring which were considered to be mutations as band sharing was otherwise far greater than expected. From Burke *et al.* 1989.

from the mother and half from the father. In this example, all bands in all offspring can be traced to either the male or the female that were resident on the territory (except for two mutant bands—see figure legend). We can be almost certain, therefore, that they were the mother and father of all the chicks. For example, with a mean of 10 paternally derived bands in each chick, the probability of another, unrelated male sharing all these bands was calculated to be less than one in a million. Even a close relative, such as a brother or father of the resident male, would have a probability of less than one in ten thousand of sharing all these bands (Burke *et al.* 1989). Although I have shown in Chapter 3 that close relatives are not involved in dunnock mating systems, this demonstrates that fingerprinting is so powerful that it could distinguish paternity even in species where close relatives were sharing matings.

7.4 Mating behaviour and paternity

In 1988, Ben Hatchwell and I made detailed behavioural observations to compare with the fingerprinting results. We observed ten monogamous pairs during the mating period. The males guarded their females closely, spending on average 81.1% of their time within 10m (range 62.1–97.3%) and intruders gained exclusive access (defined as the presence of only one male within 10m) in only three of the ten cases, a total of just 3 minutes in 23.98 hours of observation (mean of 0.17% observation time per female).

We observed 24 cases where alpha and beta males competed for matings. In 8 of the 11 cases of polyandry and 10 out of the 13 cases of polygynandry, the beta male was seen to enjoy at least some period of exclusive access to the female. The mean copulation rate during periods of exclusive access was no different for alpha males (mean 1.98 times per hour) and beta males (mean 2.15 per hour),[1] and as described in the previous chapter there was no variation in the proportion of exclusive access gained by alpha and beta at different stages of the mating period.[2] Our measures of exclusive access time, therefore, gave a good score of a male's mating success. As in monogamy, intruding males gained very little exclusive access, in only three out of the 24 cases, totalling just 5 minutes in 112.7 hours of observation (mean of 0.05% observation time per female).

As expected from these mating observations, the fingerprinting showed that intruding males gained little success. Of the 133 young from 45 broods, all but one were fathered by a male resident on the territory. The exception was a case of polyandry, where one chick from a brood of four was fathered by an alpha male from a neighbouring territory. This offspring's fingerprint consisted of 25 bands, 14 of which it shared with the female on the territory, who was therefore its mother. All 11 of the paternal bands were present in the neighbouring alpha male. We can be sure, therefore, that this neighbour was the cuckolder.

Table 7.2. Analysis of paternity by DNA fingerprinting (from Burke *et al.* 1989).

Mating system	% Total young (n) fathered by		Mean % paternity per brood (n)	
	Alpha male	Beta male	Alpha male	Beta male
Monogamy (1♂ 1♀)	100	− (49)	100	− (15)
Polyandry (2♂ 1♀)	52.9	44.1 (34)	49.2	48.5 (11)
Polygynandry (2♂ 2♀ and 2♂ 3♀)	72.0	28.0 (50)	68.4	31.6 (19)

Details for each nest:
  (a) *Monogamy*. No. chicks/total, fathered by resident monogamous male.
All young in 6 broods of 4, 7 broods of 3 and 2 broods of 2.
  (b) *Polyandry*. No. chicks/total, fathered by alpha male: 4/4, 3/4*, 3/3, 2/2, 2/4,
1/4, 1/4, 1/3, 1/3, 0/2, 0/1 (* one chick fathered by a neighbouring male;
all others fathered by beta male).
  (c) *Polygynandry*. No. chicks/total, fathered by alpha male: All young in 5 broods
of 3, 3 broods of 2, and 1 brood of 1, and 3/4, 3/4, 2/4, 2/4, 2/3, 1/3, 1/2, 0/2,
0/1, 0/1 (all others fathered by beta male).

In all cases, maternity was assigned to the resident female on the territory, so there was no intraspecific brood parasitism. This again accords with behavioural observations; we never saw neighbouring females on a nest and never recorded more than one egg being laid in a nest on any one day.

All chicks from monogamous mating systems were sired by the resident monogamous male, while in polyandry and polygynandry paternity was often shared between alpha and beta males, just as we would predict from the observations that matings were often shared (Table 7.2). Of the 30 broods raised on territories where there were alpha and beta males, the alpha male fathered the whole brood in 12 cases, the beta male did so in 5 cases and there was shared paternity in 13 cases (12 shared between alpha and beta; 1 shared between alpha and the neighbouring male). There was no difference in the proportion of young sired by the two males in polyandry and polygynandry.[3] Overall, alpha males fathered 54 of the 84 young (64.3%) and the beta males fathered 29 (34.5%). The significantly greater success of the alpha male[4] is exactly what we would expect from his greater share of the mating access (Chapter 6). Details for each brood are given at the bottom of Table 7.2 and Fig. 7.2 gives two examples of fingerprints from polyandry, one where the alpha male sired the whole brood and one where alpha and beta males shared paternity.

Table 7.3(a) analyses polyandry and polygynandry in more detail, comparing paternity in cases where a male was seen to gain mating access with cases where

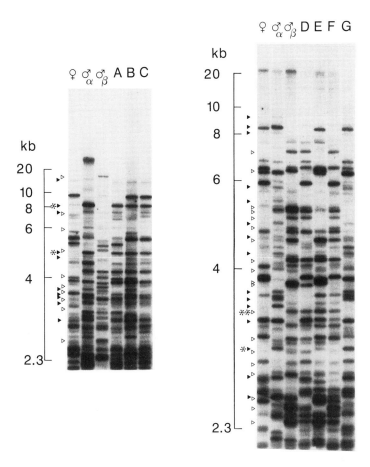

Fig. 7.2 Fingerprints from two broods on territories where there were two males and a female (polyandry). The first three columns give the fingerprints of the three adults, followed by those of the young. Paternity can be assigned by the presence/absence of diagnostic (paternal male-specific) bands in nestlings. Markers indicate diagnostic alpha (►) and beta (▷) male bands. In the left hand example, all the bands in all three offspring A, B and C can be traced to either the female or the alpha male, so the alpha male sired all young. For example, the two unique alpha male bands marked (*) turn up in all three offspring. No unique beta male band appears in any of the young. In the right hand example, by contrast, there is mixed paternity, with the beta male siring offspring D, E and F and the alpha male siring offspring G. For example, the unique beta male band marked (**) is present in all three offspring D, E, F while the unique alpha male band (*) is present in offspring G. From Burke *et al.* 1989.

Table 7.3. Linking observations on the behaviour of alpha and beta males in polyandry and polygynandry to their paternity, assessed by DNA fingerprinting. Chick feeding by males recorded for 29 of the 30 broods in Table 7.2. There were observations on matings for 18 of these cases (from Burke *et al.* 1989).

Behavioural observations	No. broods where male had some paternity	
	Alpha male	Beta male
(a) *Mating access to female*		
Had some access	13/17	10/12
Had no access	0/1	0/6
(b) *Feeding the young*		
Male fed young	23/25	14/18
Male did not feed	1/4	3/11

Beta males more likely to have paternity when seen to gain mating access ($P = 0.005$). Sample size for alpha males too small to test. Both alpha ($P = 0.011$) and beta males ($P = 0.011$) more likely to feed the young where they have paternity than where they do not. (Significance levels refer to Fisher exact tests.)

we did not see him gain access. These data confirm that our behavioural observations were sufficient to give a good measure of mating success. In none of the seven cases where we scored a male as having 'no access', did he sire any young. Six of these were where a beta male had been exluded by an alpha male and one was a case of polygynandry where an alpha male guarded only one of two synchronously laying females and ignored the other. By contrast, both alpha and beta males each had a good chance of gaining some paternity of the brood in cases where they succeeded in mating with the female. Furthermore, as shown later in Chapter 12 (see Fig. 12.2), the males' share of paternity was correlated with their share of mating access.

For 20 broods where both males in polyandry and polygynandry had access to the female during the mating period, or both fed the young (and so, by implication, both had had access—see Table 7.1), the mean paternity split per brood was 55% alpha:45% beta (total of 58 young, 56.9% sired by the alpha male). The mean percentage of exclusive mating access enjoyed by the alpha male was 61.3% ($n = 12$ of these cases where we had detailed mating observations). Overall, therefore, share of mating access gives a remarkably good prediction of share of paternity.

7.5 Paternity and parental care

A male's decision to feed chicks based on whether he had gained mating access with their mother (Table 7.1) therefore makes good sense. Table 7.3(b) shows that by using this simple rule, a male is likely to provision broods where he has paternity.

There was no indication that males could recognise their own offspring. Two results argued against this. First, Table 7.3(b) shows that males sometimes made mistakes. In two cases the alpha male fed the brood even though he had no paternity and in four cases the beta male did so. In two of these cases both males helped to feed a single chick—obviously they could not both be the father!

It could be argued that there is still chick recognition going on, but it is not perfect. However, a second piece of evidence allows us to dismiss this possibility. When the young leave the nest they are often divided among the parents, with each adult taking sole care of part of the brood and feeding them for a further two weeks until they become independent (Byle 1990). There was no tendency in such brood division for males to pick out their own young for care. We observed 28 young as dependent fledglings, from 12 broods where both an alpha and a beta male had helped to feed the nestlings. Eleven young were sired by the alpha male, of which five were cared for by their father and six by the beta male, while seventeen were sired by the beta male, of which nine were cared for by their father and eight by the alpha male.

A male with two fledglings. Males help to feed the young only if they copulated with the mother earlier on in the mating period and DNA fingerprinting shows that they are more likely to help in cases where they have paternity. However, polyandrous males do not discriminate in favour of their own sired young.

The apparent inability of dunnocks to recognise their own young is not surprising, given that they will feed young cuckoos *Cuculus canorus* and also the young of other species introduced into their nests (Chapter 13). Nevertheless, a rule of chick feeding in relation to access to the female during the mating period, which is a good predictor of paternity, leads the males to behave adaptively. I shall show later, in Chapter 12, that polyandrous males do not simply adopt a 'feed versus do not feed' rule based on mating access, but rather vary their parental effort in relation to their share of mating access. A polyandrous male therefore puts greater effort into caring for broods where he is likely to have a greater share of the paternity.

7.6 A puzzle—why no paternity marker?

Why doesn't a male dunnock pass on a paternity marker to his offspring, so that he can restrict his provisioning to his own young? There is good evidence that parent birds can discriminate their own young by recognising their distinctive calls, especially in colonial species where the young are most likely to intermingle with those of neighbours. Furthermore, the chicks of colonial species tend to give more individually distinctive calls, as if to aid parental discrimination (Beecher 1982).

In all these cases, however, the parents have to learn their young's signature calls and usually do so shortly before the young are likely to get mixed up with others. For example, young bank swallows (= sand martins *Riparia riparia*) begin to give distinctive calls at around 15 days of age, when they come to the burrow entrance to be fed. They leave the nest three or four days later, and their parents use the distinctive calls of their own young to locate them in the colony. Michael Beecher and his colleagues (1981*a*,*b*) have shown that parents will learn the calls of cross fostered chicks just as readily as those of their own young, and will care for them provided they are placed in the nest burrow before *c*. 15 days of age. Young cross fostered after this age are rejected. When some cross fostered young happened to return, after fledging, to their own home nest they were rejected by their true parents!

These experiments show that the swallows first use an indirect cue to recognise their young, namely 'any chick in my nest', and then learn the signatures of these young for discrimination away from the nest later on. Dunnocks too, apparently use this same indirect cue, which makes them susceptible to brood parasitism by cuckoos (Chapter 13), though males add a qualifier so the cue becomes 'any chick in the nest, provided I have mated with the female attending that nest'. Some rodents use similar indirect cues for offspring recognition, with males tolerating pups provided they copulated with the mother but killing the pups if they did not copulate with her (Mallory and Brooks 1978). Males show the same responses even when pups are exchanged at birth so that none of them

are their own, so males recognise their 'own' offspring by past sexual association with the mother rather than by characteristics of the young themselves
(Labov 1980). To discriminate their own young directly, male rodents and
dunnocks would need a more refined cue based on a genetic marker. Is this, in
principle, a possibility?

There is some evidence that animals can indeed use genetic markers to
discriminate close kin from more distant kin, even when these are raised in a
common environment such as the same nest (Waldman 1988). For example,
honey bee *Apis mellifera* workers can discriminate between larvae which are
full-sisters (fathered by the same male) and half-sisters (fathered by a different
male) and prefer to rear the former to become new queens (Page *et al.* 1989).
In Belding's ground squirrels *Spermophilus beldingi*, individuals behave more
altruistically towards full-sister litter mates than towards half-sister litter mates
(Holmes and Sherman 1982). In these cases it is likely that discrimination is
based on genetic odour labels, with individuals perhaps comparing the similarity
of their own label with those of their nest mates to recognise close relatives, a
process Holmes and Sherman have called 'phenotype matching'.

Birds do not have the array of odour cues available to insects and mammals,
so one answer to the problem of why there is no paternity marker in dunnocks
is simply that they may be unable to pass on a distinctive label to their offspring
and so are forced to rely on cruder, indirect, cues to paternity instead. However,
this may not be the whole story. Patrick Bateson (1982) has shown that female
quail *Coturnix coturnix japonica* raised with siblings prefer to mate with first
cousins rather than third cousins or unrelated individuals. In this case the
discriminations were probably based on the visual experience they had had with
the plumage of their close kin (ten Cate and Bateson 1989). However, quail
reared in mixed groups of kin and non-kin prefer to associate with siblings over
nonsiblings (Waldman and Bateson 1989). The cues involved are not known,
but are likely to involve plumage characteristics together, perhaps, with
vocalisations and kin recognition may involve a 'self matching' process. This
example suggests, perhaps, that refined kin discrimination based on genetic
markers may indeed be possible in birds. Why, then, don't dunnocks show this?

In some cases it may pay an individual to advertise its kinship because it gains
a benefit, as do others who act on this information. In the quail, kin recognition
may have evolved to facilitiate mate choice with both males and females gaining
an advantage by an optimal degree of outbreeding (Bateson 1983). In other
cases, however, it clearly pays individuals to hide their kinship. An egg dumped
in a nest by a brood parasite, and the chick hatching from that egg, may
be discriminated against by the hosts if their origin is advertised (Chapter 13).
Similarly, a chick fathered by a neighbouring male would be discriminated
against by the resident male if it advertised its paternity.

In dunnock broods with mixed alpha and beta male paternity it might likewise
pay chicks to hide their paternity. If they exhibited a paternity marker, they

would presumably be fed only by their father. If they hid their paternity, then they may be fed by both males and so gain even more nourishment (see Chapter 8). The female, who is mother of the whole brood, would likewise benefit if two males cared and so genes which suppressed paternity markers may be favoured. The general point, as with many other situations in the lives of dunnocks, is that we should not expect traits to evolve simply because they are of benefit to one party. Conflicting interests need to be considered and a mathematical model to explore the outcomes of this conflict over paternity advertisement would be very worthwhile. However, it is at least an entertaining possibility that the absence of direct recognition of their own offspring by male dunnocks is a case of males losing out in a conflict of interest with the female and the chicks (see also Beecher 1988).

7.7 Summary

Males fed the young only if they gained mating access to the female earlier on during the mating period. Thus, in cases where both an alpha and a beta male mated with the female, the young were often provisioned by two males and a female, whereas in cases where only the alpha male mated, only he and the female fed the chicks. Neighbours never helped to feed a brood, even in the few cases they gained mating access. It therefore pays females, as observed, to encourage matings from beta males, but not neighbours, to gain increased male help with chick feeding.

DNA fingerprinting showed that mating access was a good predictor of paternity. Over 99% of the young were fathered by resident males. Where both alpha and beta males mated with a female, paternity of the brood was often shared, with alpha males fathering on average 55% of the young and beta males 45%. A beta male's paternity share increased with his share of mating access. There were no cases of intraspecific brood parasitism; in all cases maternity was assigned to the resident female on the territory.

There was no indication that males could recognise their own offspring. If they mated with the female they fed the young even in cases where they had no paternity. When the young fledged and the brood was divided among the adults for care to independence, there was no tendency for males to prefer their own young. Males therefore used an indirect cue for paternity, based on mating success. On average, this led them to provision broods where they had some paternity.

Why don't males pass on paternity markers? One possibility is that they are unable to do so. Alternatively males may have lost out in an evolutionary conflict with the females and the young, to whom it may be advantageous to suppress paternity advertisement.

STATISTICAL ANALYSIS

1. Sixteen 'pairs' of alpha and beta males observed. Wilcoxon matched pairs test, NS.

2. For the 1988 samples, described here, the percentage of time alpha males gained exclusive access did not vary from the period 8 days before laying to incubation (Kruskal-Wallis 1-way ANOVA, $\chi^2_9 = 8.99$, $P > 0.30$) nor did that of beta males ($\chi^2_9 = 0.521$, $P > 0.90$).

3. Comparing number of young fathered by alpha versus beta males in polyandry and polygynandry in Table 7.2, $\chi^2_1 = 1.952$, $P > 0.10$. Comparing proportion of young fathered by alpha males in each brood, polyandry versus polygynandry, Mann-Whitney U-test, $P > 0.10$, 2-tailed.

4. Alpha males fathered significantly more than half of the 84 young, $\chi^2_1 = 6.857$, $P < 0.01$.

8

Reproductive output from the different mating systems

8.1 Introduction

The aim of this chapter and the next one is to assess male and female reproductive success in the various mating systems. It is clear from the behavioural observations that there are conflicts of interest both during the setting up of breeding territories and during the mating period. Can we make sense of these given the assumption that individuals are expected to behave in ways that promote their own reproductive success? Can the outcomes of these conflicts help us to understand the dunnock's variable mating system?

In this chapter I shall consider the factors which influence the reproductive output of the various mating systems. In Chapter 9 I shall then use these results to calculate male and female reproductive success in the different combinations. The analysis is based largely on data collected in the four breeding seasons from 1981 to 1984 when I simply observed natural variation in reproductive success. In the following seasons there were extensive experimental manipulations, so these have been included only in cases where they have illuminated the causes of the natural variation.

8.2 Provisioning of nestlings

The young are fed on billfuls of small invertebrates especially spiders, beetles, flies, aphids, collembola and small insect larvae, with only the occasional large items, such as earthworms and Lepidoptera caterpillars (Bishton 1985; Tomek 1988). Faecal analysis showed that the young in the Garden were fed on a wide variety of prey with the adults apparently being generalist foragers and picking up any small invertebrate they came across while they were hopping about under the hedgerows and bushes. However, adults were also quick to exploit temporary gluts of particular prey types, such as the emergences of bibionid and chironomid flies in spring.

The number of adults who helped to feed the nestlings varied with the mating system, as follows.

(a) In *monogamy* (one male with one female), the female had the full-time help of one male.

(b) In *polygyny* (one male with two females), a female had no male help, part-time help or full-time help from her male depending on the degree of reproductive synchrony between the two females (see Chapter 11 for details). If one of the females was incubating, the male was free to help the other full-time in the provisioning of young. However, if the brood overlapped either the mating period of the other female or the nestling period of her brood then his help was often reduced. On average, a polygynous female could only expect part-time help from her male.

(c) In *polyandry* (two males with one female) the female gained only one male's full-time help (the alpha) if the alpha male monopolised all the matings but two males' full-time help (alpha and beta) if both the males gained mating access (see Chapter 7).

(d) In *polygynandry* (two males, usually with two females), a female gained either no male help, one male part-time or full-time help, or two male part-time or full-time help, depending on how the two males shared matings and also on the reproductive synchrony of the two females. I shall postpone discussion of these various complications until Chapter 11. For the moment it is sufficient to note that, on average, a polygynandrous female can expect only the equivalent of one male's full-time help.

Figure 8.1 shows that the provisioning rate to the brood increased with brood size, but there were marked differences between mating systems reflecting the differences in male contribution. Nestlings were fed at a greater rate in monogamy, where one male gave full-time help, than in polygyny, where one male gave only part-time help (Fig. 8.1a). Likewise, nestlings were fed at a greater rate in polyandry where two males gave full-time help, than in poly-gynandry where two males shared their help among several females (Fig. 8.1b). Comparing all systems together (Fig. 8.1c), nestlings were fed at the greatest rate with polyandry (two males full-time), less with monogamy and poly-gynandry (one male full-time or two males part-time) and least with polygyny (one male part-time).

Do these differences really reflect differences in amount of male help or could there be confounding factors? For example, we saw in Chapter 4 that poly-androus females have larger territories than monogamous females; perhaps this difference contributes to the greater provisioning rate. The data in Fig. 8.2 argue against this possibility. First, there was no difference between the rates at which

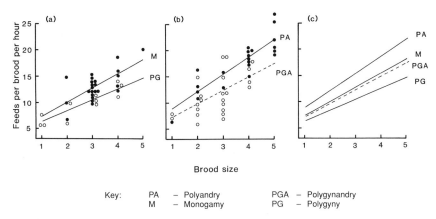

Fig. 8.1 Mean provisioning rate to nests per hour for different mating systems and for broods of different size. Each point refers to a different nest, observed when the chicks were 7–11 days old. From Davies 1986. (a) Where the female has full-time help from a male in monogamy (•, $n = 24$; $y = 2.678x + 4.632$), provisioning rate is greater than where she has only the part-time help of a polygynous male (○, $n = 11$; $y = 2.049x + 4.419$). Analysis of covariance; $F_{1,31}$ slope = 0.654, NS; $F_{1,32}$ elevation = 7.331, $P < 0.02$. (b) Where the female has the full-time help of two males in polyandry (•, $n = 21$; $y = 3.337x + 5.469$), the provisioning rate is greater than in polygynandry (○, $n = 23$; $y = 2.569x + 4.751$) where she has, on average, the equivalent of only two males' part-time help ($F_{1,40}$ slope = 0.753, NS; $F_{1,41}$ elevation = 9.920, $P < 0.01$). (c) The four regression lines together for comparison. The top line, cooperative polyandry, is significantly higher in elevation than the monogamy line ($F_{1,42}$ elevation = 18.94, $P < 0.001$) but does not differ in slope ($F_{1,41} = 0.709$, NS). There is no difference between monogamy and polygynandry.

broods were fed by monogamous pairs and on polyandrous territories where only the alpha male and female fed the brood (Fig. 8.2b). Both rates were significantly less than cases of polyandry where both males helped. Second, some trios became pairs when either the alpha or beta male died. The provisioning rates at these nests was no different from other monogamous pairs (Fig. 8.2a). Both results suggest that it is the number of provisioning adults which determines feeding rate rather than any differences in bird quality or territory quality between pairs and trios.

8.3 Provisioning rate and nestling weight

A higher provisioning rate does not necessarily indicate a greater amount of food. Billfuls of prey vary in size and parent birds tend to bring smaller loads

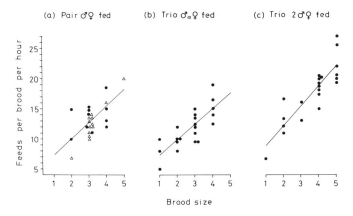

Fig. 8.2 The difference in provisioning rate was related to the number of adults who fed the brood rather than differences between pair (monogamy) and trio (polyandry) territories. (a) In cases where one male of a trio had died, thus leaving a pair to feed the brood (△), the feeding rates followed the same relationship with brood size as in other monogamous pairs (●). (b) In cases of polyandry where only the alpha male mated and fed the brood ($y = 2.556x + 4.804$; $n = 21$) provisioning rates were significantly less than in polyandry where both males mated and fed (c), $F_{1,38}$ slope $= 1.55$, NS; $F_{1,39}$ elevation $= 17.19$, $P < 0.001$. However, they were no different from monogamy, where also just one male and female fed the brood (a), $F_{1,41}$ slope $= 0.025$, NS; $F_{1,42}$ elevation $= 0.092$, NS. From Houston and Davies 1985.

when they are collecting food nearer the nest (Orians and Pearson 1979; Kacelnik 1984). A higher rate could, therefore, reflect lots of trips but with smaller loads. It was impossible to assess load size by observation, but there was no significant variation in the distances from the nest that males and females collected food in the various mating systems (Hatchwell and Davies 1990). Furthermore, mean chick weight per brood increased significantly with provisioning rate.[1] Thus, in dunnocks, a higher provisioning rate did apparently indicate increased food intake.

During 1981–1984, a total of 154 broods were weighed when the nestlings were 6 days of age (hatching day = day 0). There was no significant variation in nestling weight between years, or with season, or with brood size (Davies 1986). Nestling weight did, however, vary markedly with the mating system. Chicks were heavier, and less likely to die in the nest from starvation, when more adults fed them (Table 8.1).

Do differences in the number of providers cause these differences between mating systems or are other factors involved, such as territory quality? Two sources of evidence suggest that it is indeed the number of providers which is

Table 8.1. Influence of number of adults who feed the young in the different mating systems on nestling weight (day 6) and starvation in the nest (from Davies 1986).

Mating system	No. adults who provisioned the brood	Mean nestling weight per brood (g) ± 1 SD	(n)	% nests where one or more nestlings starved to death
Polygyny	♀ + 1♂ part-time	10.96 ± 3.02	(14)	57
Monogamy	♀ + 1♂ full-time	12.30 ± 1.76	(45)	30
Polygynandry	♀ + 2♂ part-time	11.80 ± 2.45	(33)	39
Polyandry	♀ + 1♂ full-time	12.68 ± 1.72	(24)	15
	♀ + 2♂ full-time	14.06 ± 1.28	(31)	13

The variation in nestling weight is significant (Kruskal-Wallis 1-way ANOVA, $H_4 = 24.374$, $P < 0.001$) as is the variation in starvation ($\chi^2_4 = 20.32$, $P < 0.001$).

the main influence on nestling weight. First, the differences in Table 8.1 are related to differences in the number of adults who actually fed the young rather than simply the number of adults on a territory. Thus there was no difference in nestling weight between monogamous territories and polyandrous territories where just one male fed the young,[2] but both had significantly lighter chicks than on polyandrous territories where two males fed the young.[3] There was no difference between nestling weights in monogamy and polygynandry,[4] reflecting the lack of difference in provisioning rates. On territories with two males, nestling weight was higher when they provided their full-time help (polyandry) rather than their part-time help (polygynandry).[5]

Second, the results of natural removal experiments show these same effects. When a predator killed a monogamous male during incubation, the chicks suffered a decrease in weight compared to normal pairs. Likewise, when a predator killed one of the polyandrous males during incubation, and the chicks were then fed by just one male and a female, they weighed less than where two males and a female fed them (Fig. 8.3).

8.4 Nestling weight and survival

After the young left the nest they were fed by their parents for a further 2–3 weeks. The young were scored as independent if they had either left their natal territory or had survived to 20 days after leaving the nest. Soon after inde-

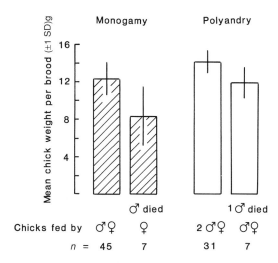

Fig. 8.3 Natural removals, caused by predators, confirm that nestling weight (day 6) is influenced by the number of adults who feed the brood. When the male dies in monogamy (hatched columns), leaving the female to provision alone, mean nestling weight is reduced (t_{51} = 2.971, P < 0.01, 2-tailed). Likewise, when one male dies in polyandry (open columns), leaving one male and the female to feed the brood, nestling weight is less than when two males help to provision (t_{37} = 3.071, P < 0.01, 2-tailed).

pendence many of them dispersed outside the study area and so the number actually seen as independent young will underestimate true survival.

Of a total of 453 nestlings ringed and weighed during 1981–1984, 229 (50.6%) were seen as independent young. Of the remaining 224 individuals, 41 died from starvation in the nest, 32 were taken from the nest by predators (mainly cats, grey squirrels and carrion crows), 17 were known to have died during the dependent fledgling period and 134 disappeared, either having died or emigrated.

Does nestling weight have an influence on chances of survival? With so many chicks 'disappearing' at around the time of independence, and so with their fates unknown, this is a difficult question to answer. As a first attempt I shall make the extreme assumption that only the 229 young actually seen to be independent survived, and all the others died. Under this assumption, survival chances for a given nestling weight (at day 6) did not vary across the 4 years.[6] Combining all 4 years' data, survival chances for a given nestling weight did not vary across the different mating systems.[7] Considering all mating systems together, survival chances were strongly dependent on nestling weight with heavier chicks on day 6 in the nest being more likely to be seen as independent young (Fig. 8.4a).

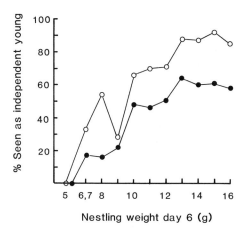

Fig. 8.4 The influence of nestling weight on the chances of survival to independence. (a) ●, proportion seen as independent young from the total of all nestlings ringed at 6 days of age. Spearman rank correlation, $r_s = 0.900$, $P < 0.01$. (b) ○, proportion seen as independent young from the total of those known to have either definitely survived or to have died, $r_s = 0.910$, $P < 0.01$. From Davies 1986.

This result could be confounded if, among the disappearing young, lighter chicks were more likely to emigrate rather than die. If this happened then the mortality among the lower weight classes would be overestimated. However, even if the analysis excludes all the disappearing young and considers only the 319 nestlings who were known to have either definitely survived to independence or to have died before independence, then the same trend emerges, with heavier nestlings being more likely to survive (Fig. 8.4b). This result has been found in several other species (Magrath 1991).

8.5 Number of young fledged as a measure of reproductive success

It is notoriously difficult to measure reproductive success. Not only is it hard to follow chicks after they have fledged to score their survival to independence, but even after independence there may be differential survival in relation to the quality of care beforehand (Perrins 1965; Gustafsson and Sutherland 1988; J.M. Tinbergen and Boerlijst 1990). It might be thought that the ideal measure would be number of offspring surviving to breed, but even this may not suffice because an offspring's fecundity may also be influenced by the care it received as a youngster (Gustafsson and Sutherland 1988).

Given these problems, any attempt to measure dunnock reproductive success seems doomed to failure. Most of the young dispersed from the Garden soon after independence so their survival to breeding age and subsequent performance were impossible to follow. However, I shall show in this section that the simple measure of 'number of young fledged' gives at least a reasonable measure of reproductive success, and given the large differences in parental care between the different mating systems, the measure should reflect their differences in reproductive output.

The previous sections showed that nestling mortality and nestling weight varied with the number of adults who fed the brood. The key question to ask is, do the differences in reproductive success emerge by the time the young leave the nest (in which case number fledged will give a good measure of reproductive success) or is there differential mortality depending on mating system after fledging? To provide a sufficient sample size for analysis, I have considered the 723 young (from 250 broods) which fledged successfully in the breeding seasons 1981−1988. Of these, 9% remained in the Garden to breed and so were known to survive to breeding age. The remainder will include birds which dispersed to breed outside the study area and those which died after fledging and before the next breeding season.

When two males copulate with a female, both may help to feed her brood. Provisioning rate, nestling weight and number of young fledged are greatest when two males give their full-time help in a polyandrous mating system.

There were no differences between mating systems in the proportion of successful fledgers which stayed to breed,[8] nor were there any differences within polyandry and polygynandry between trio and pair-fed chicks.[9] This suggests that the differences between mating systems in chick mortality all emerge before fledging. The young who stayed to breed were also no different in their day 6 nestling weight than young who fledged successfully but did not breed in the Garden.[10] This suggests that the differences in survival with day 6 weight shown above must be largely due to mortality before fledging. After fledging, there was no hint of differential survival in relation to day 6 weight. This result supports the finding in the previous section that survival chances for a given day 6 nestling weight did not vary across the different mating systems.

The main conclusion is that there was no evidence of differential survival after fledging across the different mating systems. The number of young fledged therefore provides a reasonable measure of reproductive output from the different systems.

8.6 Failed breeding attempts: interference and infanticide

I now consider how the number of young fledged per nest varies among the different mating combinations. A breeding attempt was defined as one in which at least one egg was laid and the analysis is based on a total of 227 breeding attempts during 1981−1984, involving 78 different males and 70 females. There was no significant variation across the four years of study[11] and so these have been combined to give the summary in Table 8.2. Reproductive output per nest varied significantly across the different systems.[12]

The proportion of failures (0 chicks fledged) varied with the mating system,[13] being particularly high in polygynandry and in polyandry where the beta male failed to mate and so only the alpha male provisioned the brood. Table 8.3 shows the causes of these failures. There are two particular points of interest.

First, there was a high frequency of egg desertion by females in polygynandry. At least seven of these were a direct result of aggression between females involved in the same mating system, and most of the other desertions probably had the same cause. Females who were laying eggs or feeding chicks sometimes visited incubating females in the same polygynandrous system and attacked them on the nest. Incubating females also sometimes left their nests to attack another female feeding nearby. Fights were sometimes prolonged, with the two females interlocked and grappling on the ground. During these fights between polygynandrous females, the males tried to intervene and keep their squabbling females apart; a male often went between the two females and chased each, in turn, back towards its own territory. However, females sometimes

Table 8.2. The number of young fledged per nest in the different mating systems (1981–1984) (data from Davies 1986).

Mating system (n = no. nests)	No. young fledged per nest						Mean	
	0	1	2	3	4	5	All attempts	Successful attempts only
Polygyny (n = 21)	8	4	4	2	3	–	1.43	2.31
Monogamy (n = 62)	21	2	6	25	8	–	1.95	2.95
Polygynandry* (n = 47)	22	4	9	9	3	–	1.29	2.44
Polyandry Only alpha ♂ fed (n = 38)	22	2	2	8	4	–	1.21	2.88
Alpha and beta ♂ ♂ fed (n = 39)	6	1	7	5	15	5	2.95	3.48

* Polygynandry refers to 2♂ 2♀.

Table 8.3. Causes of failed breeding attempts (0 chicks fledged) in the different mating combinations (1981–1984).

Cause of failure	Number of cases				
	Polygyny	Monogamy	Polygynandry*	Polyandry	
				α ♂ mated	α & β ♂ ♂ mated
Female killed	1	7	2	3	1
Nest depredated	2	7	7	8	4
All young starved to death	3	2	3	2	–
Eggs deserted	1	4	13	1	1
Eggs cracked/young chicks removed	–	–	4	6	–
Unknown	1	1	1	2	–

* Includes 2♂ 3♀.

deserted after these fights and then begun another nest, usually further away from the other female.

The second point of interest in Table 8.3 is the high frequency of cracked eggs or removal of young chicks, often one by one, in cases of polygynandry and polyandry where the beta male failed to mate (or did not feed the chicks, and so by implication had failed to mate—see Chapter 7). These acts of interference were not accompanied by any disturbance to the nest itself, which usually occurred when a predator was the cause. In two cases the eggs had a little hole in the shell, which looked to be the result of a peck. Beta males who failed to copulate harassed incubating females on the nest and I strongly suspect that they were the cause of these failures. An advantage to a beta male from behaving in this way is that if the eggs or chicks are destroyed then the female begins a replacement clutch within a week or two, so this hastens the day that the beta male has another chance to copulate and so rear his own young. From his point of view, this is clearly better than waiting on the sideline while the alpha male and female rear the brood, which would involve a wait of four to six weeks to the next breeding attempt. We may wonder, then, why beta males do not always interfere in this way. A likely answer is that alpha males are often able to prevent them. Alpha males were seen chasing beta males away from the nest vicinity and beta males may have risked injury to themselves which could outweigh the costs of a longer wait to the next mating chance.

Although there was no direct evidence for beta male destruction of eggs or nestlings, the following data offer indirect support for the interpretation above. Figure 8.5 shows that the probability of a female having another breeding attempt declined sharply towards the end of June. If the advantage to beta males from interference is that females begin a replacement clutch, then beta males would gain little from interference towards the end of the breeding season because the chances that the female will lay again are small. Of the ten cases of suspected beta male interference (Table 8.3), one was in March, three were in April and six were in May. There were no cases after the time a female was unlikely to re-lay. In fact, towards the end of the season beta males who failed to mate often remained alone and began their autumn moult.

Considering the seasonal occurrence of all failed breeding attempts, it is remarkable that the only mating combination to show a significant decline in failure rate at the end of the season, when the female is unlikely to re-lay, is polyandry where only the alpha male mated and fed the brood (Table 8.4). This provides indirect evidence for a change in beta male tactics at the end of the season. It is especially interesting that the frequency of failure remains high throughout the season in polygynandrous combinations, where much of the interference is caused by females. Interference between females is predicted to continue throughout the season because one advantage from causing the other female to desert is that the aggressor can then claim increased parental care from the males (Chapter 11). This advantage would occur at the end of the season just as much as at the beginning.

Infanticide in animals may occur in various circumstances (Hrdy 1979; Sherman 1981). In mammals it has long been recognised that males who fail to copulate with a female may kill her young. As a result, the female comes into

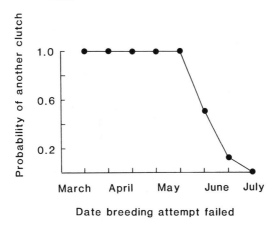

Fig. 8.5 Seasonal change in the probability that a female laid another clutch if her present breeding attempt failed. The season is divided into eight periods. Sample sizes from left to right are; 4, 10, 17, 11, 7, 8, 8, 3. From Davies 1986.

Table 8.4.   Seasonal occurrence of failed breeding attempts in the different mating systems (1981–1984). The season is divided into March–May, when females were certain to have another breeding attempt if they failed, and June–July, when they were less likely to lay again (see Fig. 8.5) (from Davies 1986).

Mating system	% breeding attempts failed (n)		Significance of difference
	March–May	June–July	
Polygyny	42 (12)	33 (9)	NS
Monogamy	36 (42)	30 (20)	NS
Polygynandry	48 (46)	47 (17)	NS
Polyandry			
Only alpha male			
mated/fed	74 (27)	18 (11)	$P < 0.01$
Both males mated/fed	19 (26)	8 (13)	NS

Significance levels from χ^2_1 tests. Polygynandry includes 2♂ 3♀.

oestrus again sooner, and so infanticide hastens the day that the male has a chance to sire his own offspring (e.g., lions, Bertram 1976, Packer and Pusey 1983; langurs, Hrdy 1977; rodents, Mallory and Brooks 1978, Labov 1980). The first direct observations of this kind of infanticide by male birds was by J.R. Crook and Shields (1985) for swallows *Hirundo rustica*. Since then there has been convincing experimental evidence for other species that males are likely to destroy eggs or kill young chicks when they have failed to copulate with the female. When Robertson and Stutchbury (1988) removed resident male tree swallows *Tachycineta bicolor* shortly after the eggs hatched, new males quickly took over and often killed the chicks. The new males then either mated with the female for a replacement clutch or took over the nest site with another female. Walter Koenig (1990) has shown that when dominant male acorn woodpeckers are removed experimentally from polygynandrous breeding groups during the egg laying period, so that they fail to father any of the brood, they destroy the clutch on their return to the territory, thus forcing a re-nest. In control experiments, where dominant males were removed during incubation, after all the mating was over, they never destroyed the clutch. This shows that it was their failure to gain copulations, not the removal experience itself, which induced clutch destruction.

Ben Hatchwell and I performed similar removal experiments in dunnocks, but our removals were of short duration (3 days) and were designed to test how reduction of paternity influenced parental effort (see Chapter 12). We found no evidence that infanticide resulted from these removals, but none of our removals were of sufficient duration to ensure that a male failed to copulate at all during the mating period. Further experiments are needed but I feel that the indirect evidence for male infanticide in dunnocks is quite convincing.

Veiga (1990a) has found evidence for infanticide by both male and female house sparrows *Passer domesticus*. Males who had recently lost their mates destroyed eggs or killed and removed young chicks from the nests of neighbouring females, and subsequently bred with those females. Females who shared a male in polygyny were also seen destroying the eggs or chicks of other females in the same mating system, and apparently thereby gained increased help from the male with parental care. These two contexts for infanticide, with differing advantages to males and females, are exactly the same as those which occur in the dunnocks.

8.7 Successful breeding attempts: parental care and fledging success

The difference in reproductive success with mating systems in Table 8.2 is not only due to differences in failure rates because even considering just successful breeding attempts (those which produced at least one fledged young) there was still significant variation.[14] The following conclusions emerge from an analysis to see where these differences lie, by making pair-wise comparisons between different mating systems.

(a) In polyandry, reproductive success was greater where both males mated and fed the brood than where only the alpha male did so, considering either all breeding attempts[15] or only successful attempts.[16] Greater success in cases where both males mated was therefore due to both a lower failure rate (no beta male interference) and better chick production in successful attempts (two male care rather than one male care).

(b) Polyandrous trios where both males helped were more successful than monogamous pairs, both considering all breeding attempts[17] and only successful attempts,[18] reflecting better chick provisioning when three adults fed the chicks as opposed to two.

(c) Polyandrous trios where only the alpha male mated and fed the brood were less successful than monogamous pairs when all attempts are considered,[19] but not when only successful attempts are considered.[20] The difference is therefore due entirely to the greater failure rate of trios where only the alpha male mated, attributable at least in part to beta male interference. For successful attempts, the lack of difference is exactly what would be expected if the number of adults provisioning a brood was the main determinant of reproductive success; in both monogamous pairs and polyandrous trios where only the alpha male mated, one male and one female fed the young.

(d) Comparing all polyandrous trios (including cases where both males mated and only one male did so) with monogamous pairs, there was no significant difference in reproductive success.[21] This is an important result, to which we shall return in Chapter 9 when considering whether it would pay a female to try and gain two males rather than one.

(e) A polygynous female fledged fewer young per nest than a monogamous female, but the difference was not significant.[22] Given the lower provisioning rate, lower nestling weight and greater chick starvation in polygyny, reproductive success would be expected to be lower and the lack of statistical significance may reflect the small sample size for this mating system, which was not common.

(f) Polygynandrous breeding attempts were less successful than polyandry where both males helped, both considering all attempts[23] and just successful attempts,[24] as expected from the better provisioning rate, nestling weight and survival where two males helped a female full-time rather than part-time.

(g) Polygynandrous breeding attempts were less successful than those in monogamy, both for all attempts[25] and only successful attempts.[26]

In summary, these differences in reproductive output of the different mating systems are what we would expect from their differences in parental care described earlier in the chapter. Various factors are now considered which could confound this interpretation. Food supply usually varies with season and habitat and so it could be argued that if different mating combinations were predominant at different seasons, or in different habitats, then the main cause of the differences in reproductive success in Table 8.2 may be food abundance rather than the number of adults who provision the brood.

There was indeed significant seasonal variation in the number of young fledged per breeding attempt,[27] but there was no seasonal variation in the frequency of the different mating combinations.[28] Controlling for season, there was still significant variation in reproductive success across the different mating systems.[29]

There was no significant variation in reproductive success per breeding attempt across six main habitat types in the Garden.[30] Dividing the study area up in other ways, for example, considering eleven different areas of the Garden, likewise showed no significant variation in success. Within the six habitat types, there was still significant variation in reproductive success across the different mating systems.[31]

8.8 Removal experiments and matched comparisons

Although yearly variation, season and habitat can now be rejected as possible confounding variables, it is still possible that differences in reproductive success are due to differences in individual quality or territory quality. For example, perhaps the best quality females (those best able to rear chicks) are most often associated in polyandrous trios where two males feed the brood.

There are two sources of evidence against this possibility. The first comes from natural removal experiments by predators. In nine cases one of the males from a polyandrous trio was killed between nest building and chick hatching (six beta males and three alphas). Figure 8.6 shows that reproductive success in these 'experimental' cases was significantly lower than in cases where two males helped to feed the young.

The second source of evidence comes from 'matched comparisons'. As shown in Chapter 4, mating combinations changed frequently both within and between seasons due to mortality and movements. Therefore it was possible to compare the success of breeding attempts by the same female on the same territory but in different mating combinations. These matched comparisons were possible, for example, when a female sometimes bred in a trio where only the alpha male mated and fed the brood and on other occasions bred in a trio where both males mated and fed the brood. In other cases a trio became a pair when one male died or left to another territory, or a pair became a trio when another male joined them. Finally, a polyandrous female could later end up in polygynandry if her males took over another female, or a polygynandrous female could change to polyandry if another female died.

Table 8.5 summarises these matched comparisons. If a particular female had more than one breeding attempt in a certain system, then the mean reproductive

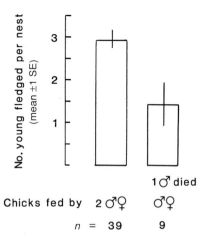

Fig. 8.6 Natural removals, caused by predators, show that when one male is removed from a polyandrous trio, so that the chicks are cared for by just one male and a female, reproductive success is less than when two males and a female care for the brood ($t_{47} = 2.452$, $P < 0.02$, 2-tailed).

success and nestling weight per brood was used in the analysis. The conclusions are very similar to those from the overall comparisons in Table 8.2. Females breeding in polyandrous trios where both males fed the young reared heavier chicks and fledged more young than when they bred as monogamous pairs or in trios where only the alpha male helped. There was no difference in chick weight or reproductive success when one male helped in polyandry versus monogamy. Polyandrous females with the full-time help of two males reared more young than polygynandrous females with, on average, only the part-time help of two males. In conclusion, the mating system influenced reproductive success even considering the same female on the same territory.

In Chapter 4, we saw how the distribution of feeding patches influenced the mating system by determining a female's territory size and hence her ease of monopolisation by males. It might be expected that these differences in territory size and quality would influence reproductive success, but there was no suggestion from any of the results above that territory quality influenced chick provisioning, nestling weight or fledging success. The effect of territory quality on reproduction was therefore at most only small, and swamped by the overriding importance of the number of adults who provisioned the young.

Why does dispersion of feeding patches influence female territory size but not the efficiency with which chicks are fed? The key to this may be the dunnock's foraging specialisation on small insects and seeds. For self-feeding, the dispersion of prey is indeed expected to influence ranging behaviour. However, for

Table 8.5. Matched comparisons of the reproductive success and nestling weights for the same female breeding on the same territory, but in different mating combinations. Data from 1981–1984 (from Davies 1986).

Mating systems compared	No. females		Mean nestling weight per brood (g)	Mean no. young fledged per nest
(i) Monogamy versus	12	(i)	12.3	2.0
(ii) Polyandry where only alpha male mated/fed brood		(ii)	12.7	1.6
(iii) Monogamy versus	7	(iii)	12.5	2.4
(iv) Polyandry where both males mated/fed brood		(iv)	14.1*	3.7*
(v) Polyandry where only alpha male mated/fed brood versus	7	(v)	12.0	1.5
(vi) Polyandry where both males mated/fed brood		(vi)	14.3*	3.6*
(vii) Polygynandry versus	6	(vii)	11.4	1.3
(viii) Polyandry where both males mated/fed brood		(viii)	13.3	3.3*

Significant differences indicated from Wilcoxon matched pairs tests, 2-tailed.
*$P < 0.05$. Other comparisons not significant.

chick feeding the major rate-limiting factor is not the distance at which food is collected from the nest, but rather the long time needed to collect a billful of prey for the young. An adult dunnock typically brings just six loads per hour to the nest (Chapter 10), having to spend long periods collecting each billful of prey. Whatever the abundance and dispersion of these small prey, collecting time will impose a severe limit on the ability of adult dunnocks to feed their young. It seems likely then that the rate of food delivery to the nest is limited not primarily by food abundance but rather by the work force that collects it. The main determinant of reproductive success is the mating system, simply because it is this which decides the work force available to collect the food.

The influence of work force on fledging success may be strongest in species, like dunnocks, which specialise on small dispersed prey with long collecting times. In species which exploit bonanzas of large prey, like caterpillars, one or two adults may easily be able to fulfil a brood's needs when prey are abundant. With small, dispersed prey, however, even when prey are abundant extra collectors may have a marked effect.

8.9 Variation in clutch size in relation to expected male help

It is clear that when more adults feed the brood, the chicks get better fed and more survive. Figure 8.7 summarises this in terms of the number of young fledging successfully from different numbers of hatchlings. Success is greatest when a female has full-time help from two males, less with help from just one and least with no male help.

Surely then, a female could estimate the amount of expected male care from the mating system she is involved in and adjust her clutch size accordingly. With increased male help, not only are larger clutches much more productive (Fig. 8.7) but the female herself also works less hard at chick feeding (Chapter 10). Without extensive experimental manipulations to examine how clutch size influences nestling and female survival under different regimes of male help, we cannot predict optimal clutch sizes (Gustafsson and Sutherland 1988; Pettifor *et al.* 1988). Nevertheless, Table 8.6 suggests that females do indeed vary clutch size in relation to the help they expect from males. This analysis is based on a larger data set including the seven breeding seasons 1981−1984 and 1988−1990 when detailed data were obtained both on reproductive success and mating behaviour (Davies and Hatchwell 1992). The results below are not affected by yearly or seasonal variation in clutch size.

In polygyny and polygynandry a female gains variable male help, from none at all to the full-time help of one or two males respectively, depending on the activities of the other female in the system. In monogamy the female is assured

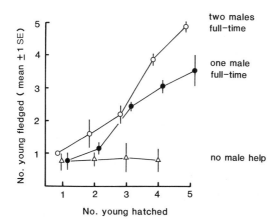

Fig. 8.7 Reproductive success in relation to the number of young hatched for different degrees of male help with parental care. From Davies and Hatchwell 1992. (For further discussion see Chapter 11.)

Table 8.6. Variation in clutch size in relation to the expected amount of male help in different mating systems (from Davies and Hatchwell 1992).

Mating system	Expected male help	No. clutches					Mean	(SE)
		2	3	4	5	(*n*)		
Polygyny (1♂ 2♀)	From none to one male full-time	–	8	10	2	(20)	3.70	(0.15)
Polygynandry (2♂ 2♀)	From none to two males full-time	5	21	58	7	(91)	3.74	(0.07)
Monogamy (1♂ 1♀)	One male full-time	2	17	43	13	(75)	3.89	(0.08)
Polyandry (2♂ 1♀)	One male full-time	1	15	18	3	(37)	3.62	(0.11)
	Two males full-time	–	7	25	16	(48)	4.19	(0.09)

ANOVA: $F_{4,266} = 4.864$, $P < 0.002$.
No difference among the first four rows ($F_{3,219} = 0.922$, $P = 0.431$), so the significant variation is due entirely to the larger clutch size in polyandry with two males full-time help, which is larger than in polyandry where just the alpha male mated ($P < 0.001$), monogamy ($P < 0.05$), polygyny ($P < 0.01$) and polygynandry ($P < 0.001$; 2-tailed t-tests).

of one male's full-time help. In polyandry where the alpha male monopolises all the matings she likewise gains just one male's help. There were no differences in clutch size between any of these mating systems in Table 8.6. By contrast, in polyandry where both alpha and beta male share the matings, the female is assured of two males' full-time help (Chapter 7), and in this case she laid a significantly larger clutch than in the other mating combinations. Matched comparisons of the clutch sizes laid by individual females involved in different mating systems for different breeding attempts showed the same result.[32]

There were not data on mating success for all the cases of polyandry in Table 8.6 and I have assumed that where one male only (the alpha) fed the brood, then only he had gained mating access and where both males fed the brood, both had gained mating access. This seems very reasonable given the clear-cut link between mating access and the feeding of chicks (Chapter 7). It should be emphasised that a beta male's decision to help is not influenced by brood size; provided he gains a share of the mating access he helps to feed even a single chick (Chapter 7). Therefore, the difference in clutch size between the two cases of polyandry in Table 8.6 is not simply an artefact of a tendency for two males to be more likely to feed larger broods. The causal link appears to be from two males gaining access to the laying of a larger clutch.

Do polyandrous females make short-term adjustments of their clutch size in relation to whether one or both males mate? We found no evidence for short-term reactions resulting from removal experiments during the mating period (Davies and Hatchwell 1992). Thus females did not suddenly reduce clutch size when their mate, or one of their mates, disappeared during the mating period. This suggests that the difference between the two cases of polyandry may arise because females can make longer term predictions about the outcome of mating competition. For example, females are better able to escape alpha male guarding and so get beta males to mate when territories are larger or have denser vegetation (Chapter 6). In natural cases the share of access by alpha and beta males does not vary through the mating period (Chapter 6), so the female will have several days over which to prepare for a larger clutch size, unlike in our experiments.

8.10 Summary

The young were fed on billfuls of small invertebrates. Nestlings were provisioned at the greatest rate in polyandry where both males mated with the female (two males' full-time help), less in polyandry where only the alpha male mated or in monogamy (both with one male full-time help), likewise less in polygynandry (on average, two males' part-time help), and at the least rate of all in polygyny (on average, only one male's part-time help). These differences in provisioning rate were reflected by differences in nestling weight and fledging success.

There was no evidence for differential survival of young after fledging, either in relation to nestling weight or mating system. Thus, the differences between mating systems emerged before fledging and the number of young fledged gave a good measure of the reproductive output per nest.

In polygynandry there was a high failure rate, mainly caused by aggression between females in the same mating combination which led to desertion. In polyandry where only the alpha male mated with the female, there was also a high failure rate. Indirect evidence suggested that beta males who failed to mate sometimes committed infanticide by pecking eggs or removing young chicks from the nest, so hastening the day they had another chance to copulate.

Removal experiments and matched comparisons of the same female breeding on the same territory, but in different mating combinations, confirmed that the main determinant of chick provisioning, nestling weight and fledging success was the mating system, because this decided the work force available to feed the young.

Females adjusted their clutch size in relation to the expected amount of male help, laying larger clutches in polyandry where both males mated and where they could therefore rely on two males' full-time help.

STATISTICAL ANALYSIS

1. Mean chick weight per brood on day 6 versus provisioning rate. (a) Broods of 2; $r_s = 0.479$, $n = 21$, $P < 0.05$. (b) Broods of 3; $r_s = 0.365$, $n = 40$, $P < 0.05$. (c) Broods of 4; $r_s = 0.667$, $n = 19$, $P < 0.01$. See Hatchwell and Davies (1990).

2. $t_{68} = 0.856$, NS.

3. Monogamy versus polyandry with two males helping, $t_{75} = 5.042$, $P < 0.001$, 2-tailed. Polyandry with one male help versus polyandry with two male help, $t_{54} = 3.231$, $P < 0.01$, 2-tailed.

4. $t_{77} = 0.990$, NS.

5. $t_{63} = 4.669$, $P < 0.001$, 2-tailed.

6. Friedman 2-way ANOVA, $\chi^2_3 = 4.912$, $P > 0.10$.

7. Considering monogamy, polyandry trio-fed, polyandry pair-fed, polygynandry and polygyny, Friedman 2-way ANOVA $\chi^2_4 = 7.218$, $P > 0.10$.

8. Comparing polygyny, monogamy, polygynandry and polyandry, there were no significant differences in either the percentage of broods fledging at least one chick which produced a chick which bred in the Garden ($\chi^2_3 = 2.07$, $P > 0.5$), nor in the percentage of chicks fledging successfully which stayed to breed in the Garden ($\chi^2_3 = 2.47$, $P > 0.30$).

9. No differences in the percentage of successful broods producing a chick which bred in Garden between trio-fed and pair-fed broods in either polygynandry ($\chi^2_1 = 0.067$) or polyandry ($\chi^2_1 = 0.031$). Likewise, no difference in the percentage of successfully fledging chicks staying to breed for trio-fed versus pair-fed chicks in either polygynandry ($\chi^2_1 = 0.009$) or polyandry ($\chi^2_1 = 0.126$).

10. Considering all young which fledged successfully, mean $\pm$ 1 SE day 6 nestling weight for young which bred in the Garden (13.06 ± 0.25, $n = 61$) was not significantly different from those which did not (12.64 ± 0.09, $n = 632$), $t_{692} = 1.581$, NS. Within broods producing a chick which stayed to breed, mean weight of chicks which bred in the Garden was not significantly different from those that did not (12.86 ± 0.22, $n = 60$), $t_{120} = 0.605$, NS.

11. Considering all mating systems, Kruskal-Wallis 1-way ANOVA, $H_3 = 5.386$, $P > 0.10$. For each mating system separately; monogamy $H_3 = 2.837$, $P > 0.30$; polyandry, trio-fed, $H_3 = 4.755$, $P > 0.10$; polyandry, pair-fed $H_3 = 1.897$, $P > 0.50$; polygynandry $H_3 = 0.443$, $P > 0.95$; polygyny $H_3 = 3.608$, $P > 0.10$.

12. Kruskal-Wallis 1-way ANOVA, $H_4 = 28.576$, $P < 0.001$.

13. $\chi^2_4 = 16.86$, $P < 0.01$.

14. Kruskal-Wallis 1-way ANOVA $H_4 = 17.779$, $P < 0.01$.

15–26. Results of pair-wise comparisons of data in Table 8.2 using Mann-Whitney U-tests, 2-tailed.

15. $P < 0.001$ 16. $P = 0.05$ 17. $P = 0.0014$ 18. $P = 0.014$

19. $P < 0.001$ 20. $P = 0.881$ 21. $P = 0.452$ 22. $P = 0.183$
23. $P < 0.001$ 24. $P < 0.001$ 25. $P < 0.05$ 26. $P < 0.05$

27. The season was divided into seven time periods; first and second halves of April, May and June and first half of July. Kruskal-Wallis 1-way ANOVA, $H_6 = 15.89$, $P < 0.02$.

28. Considering these seven time periods and the five mating systems in Table 8.2, $\chi^2_{20} = 17.95$, $P > 0.50$.

29. Friedman 2-way ANOVA, $\chi^2_3 = 12.88$, $P < 0.01$.

30. The six habitat types were: woodland, rockeries, garden, hedgerows, herbaceous border, long grass with scattered bushes. Kruskal-Wallis 1-way ANOVAs for all breeding attempts $P > 0.30$; for each mating system, $P > 0.80$ in all cases.

31. Omitting polygyny (sample size too small), Friedman 2-way ANOVA $\chi^2_3 = 13.80$, $P < 0.01$.

32. Considering 19 females where such matched comparisons were possible, their mean clutch size was significantly greater in polyandry with two male help (4.25, SE 0.13) than in other mating systems in Table 8.6 (3.84, SE 0.13; Wilcoxon matched pairs test, $P < 0.02$). See Davies and Hatchwell (1992).

9

Individual reproductive success in the various mating systems: conflicts of interest

9.1 Why measure individual reproductive success?

In this chapter I shall use the data from Chapter 8, on reproductive output from the various mating systems, to assess individual reproductive success. The analysis is based largely on work I did with Alasdair Houston (Davies and Houston 1986), combined with the results from DNA fingerprinting to measure maternity and paternity (Chapter 7). In many ways this represents the culmination of the previous eight chapters. They have raised two puzzles; why does the dunnock have such a variable mating system and why are there such intense conflicts among individuals over mate guarding and paternity?

The assumption underlying modern interpretation of animal behaviour is that individuals will adopt behaviour patterns which best promote their own reproductive success. Measuring the reproductive consequences of behaviour patterns is therefore crucial to an understanding of why individuals behave in particular ways. I shall show how measuring individual reproductive success not only helps solve the puzzle of why there are conflicts of interest in dunnock behaviour, but also provides the key to understanding why there is a variable mating system. In solving the second puzzle, therefore, we can provide an answer to the first.

9.2 Lifetime success or short-term measures?

To understand how natural selection has moulded individual behaviour patterns we sometimes need to consider how behaviour influences lifetime reproductive success because the consequences of an individual's actions are often delayed (Clutton-Brock 1988; Newton 1989). For example, it has been suggested for both Galapagos hawks, *Buteo galapagoensis*, and acorn woodpeckers, that it may pay males to cooperate in the same polyandrous or polygynandrous mating system, even though their own reproductive success per breeding attempt might be lowered through the costs of shared paternity, simply because they survive better and so have greater lifetime reproductive success than in monogamy (Faaborg *et al.* 1980; Koenig and Mumme 1987). A male's greater survival therefore offsets his lower success per breeding attempt. Another example where long-term measures are vital for understanding behaviour comes from cooperatively breeding birds, where the young remain on their natal territories and help their parents to reproduce while they wait for a breeding vacancy. In some cases, the young forgo the immediate benefits of breeding and instead wait for a vacancy on a higher quality territory. Their delay is more than compensated by the greater lifetime success on a good territory (Stacey and Ligon 1987; Komdeur 1991).

In these cases, short-term measures of breeding success, such as number of young produced per attempt, will not reflect an individual's reproductive advantages from following different courses of action. For dunnocks and other short-lived birds, however, success per attempt is likely to reflect lifetime success. Half the dunnocks only ever have one breeding season, with the chance of perhaps two or three breeding attempts, so they must make full use of every opportunity. Furthermore, there was no indication of differential survival of adults in relation to the mating system (Chapter 3, Fig. 3.5), so success per attempt is likely to reflect long-term success in the different systems.

However, it must be admitted that undetected differences in survival could be important. For example, if a polyandrous male dunnock has 4.45 young per season (half the paternity of the long season payoff in Table 9.4, see later in this chapter) compared to 5.9 young for a monogamous male, then for lifetime success to be as great, the polyandrous male would need 1.33 times as many breeding seasons. In terms of annual survival, if the monogamous male's survival chances were 65%, the polyandrous male would need 71% survival to achieve the same lifetime success. Large sample sizes are needed to detect such small differences in survival which may have large effects on lifetime reproduction.

9.3 Success per attempt or per season?

I showed in the previous chapter that 'number of young fledged per nest' gave a good measure of reproductive output per attempt from the different mating systems. It might be thought, therefore, that we could use this as our short-term measure (Table 9.1). However, there are two potential problems. The first can be dismissed but the second cannot.

(a) *Seasonal variation.* In some species, young fledged earlier in the season have a greater chance of surviving to breed (Perrins 1979). However, there was

Table 9.1. Mean number of young fledged per female for each breeding attempt and for a whole breeding season in the different mating systems. Sample sizes indicated in brackets. The first three columns come from Table 8.2 (from Davies and Houston 1986).

Mating system	% attempts successful	Mean no. young fledged per female		
		Per attempt		
		Successful attempts	All attempts	Per season
Polygyny ♂ 2♀	61.9	2.31	1.43 (21)	2.62 (8)
Monogamy ♂ ♀	66.1	2.95	1.95 (62)	5.37 (16)
Polygynandry 2♂ 2♀	53.2	2.44	1.29 (47)	3.61 (23)
Polyandry 2♂ ♀ Only alpha ♂ mated/fed brood	42.1	2.88	1.21 (38)	4.19 (16)
Both males mated/ fed brood	84.6	3.48	2.95 (39)	6.69 (13)

no such effect for dunnocks. Of the 250 broods which fledged at least one young (see Section 8.5), 63 (25%) produced one or more birds which bred in the Garden. There was no seasonal variation in the proportion of broods fledging young which produced breeding recruits,[1] nor in the proportion of fledged young remaining to breed.[2] We can assume, therefore, that broods produced at different times in the breeding season are equally valuable.

(b) *Time taken for successful and failed attempts.* The time taken for a breeding attempt was scored as the number of days from the first egg of that attempt to the first egg of the next attempt that season, in other words the time of a complete cycle. Successful attempts, those producing at least one fledged young, took on average 43.4 days (SD = 5.8, $n = 60$). There was no variation across the different mating systems in time taken to rear a successful brood,[3] nor in time taken to rear broods of different sizes,[4] nor was there any seasonal variation in time taken.[5]

On average, failed attempts (no young fledged) took only 28.2 days (SD = 8.02, $n = 44$), significantly less time than successful attempts.[6] There was no difference in the time taken for failed attempts caused by predation and those caused by suspected interference from beta males or females.[7]

The importance of the time taken per attempt is that dunnocks have a long breeding season and so have time for several attempts. Eggs are laid from the

Table 9.2. Sequences of successful (*S*) and failed (*F*) breeding attempts during a season, for 72 females breeding during 1981−1984 who survived the whole season and where all attempts were recorded. An attempt is defined as at least one egg laid (from Davies and Houston 1986). Most females had one or two successful broods per season, while a few had none and a few had three.

Sequence	No. cases	Sequence	No. cases
(a) *No successes* (*n* = 6)		(c) *Two successes* (*n* = 34)	
FFFFF	1	FFSS	2
FFFF	1	FSFS	1
FFF	2	FSS	5
FF	2	SSF	2
		SFS	4
(b) *One success* (*n* = 24)		SS	20
FFFSF	1		
FFSF	1	(d) *Three successes* (*n* = 8)	
FSFF	1	FSSS	2
FSF	2	SSS	6
FFS	3		
SFF	2		
FS	11		
SF	2		
S	1		

middle of March to the first week of July, with 90% of clutches being started in a 90 day period from 1 April to 29 June. Even if a female has several failures, because these take up less time it is still possible to have one or two successful attempts in the season (Table 9.2). The key point is that reproductive success per attempt will underestimate the success of birds which have lots of failures (Fig. 9.1). This problem is particularly acute, because there are different failure rates in the different mating systems (Table 9.1). The measure of success per attempt will therefore underestimate success in systems with high failure rates, such as polygynandry and polyandry where the beta male does not mate. I am very grateful to Kate Lessells who pointed this out to me after I had given a talk on dunnocks and made me realise that all my calculations were wrong! Measures of reproductive success clearly need to take account of the time to achieve that success (e.g., Brockmann *et al.* 1979).

A better measure, therefore, is the number of young produced by the different mating systems over the same time period, such as a whole breeding season. The last column in Table 9.1 gives these data for females who survived a whole season and who remained in the same mating combination throughout. The point

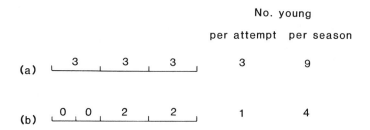

Fig. 9.1 An example to show how reproductive success per breeding attempt does not give a good indication of reproductive success per given time period when failure rates differ. (a) Three successful attempts in a season, each producing 3 young. (b) Two unsuccessful attempts, followed by two successful attempts, each producing 2 young. In terms of mean success per attempt, (a) is 3× as productive. However, because the failures in (b) each take less time than a successful attempt, success in (a) over the same time period (a season) is only 2.25× as great.

about reproductive success per attempt giving a poor indication of reproductive success in a given time can be seen by comparing polyandrous mating combinations where only the alpha male mated with those where both alpha and beta males mated and fed the brood. The former had a high failure rate and reproductive success per attempt was only 41% of that achieved by trio-fed broods. However, in terms of reproductive success over the same time period (a season) the combinations where only the alpha male mated produced 63% of the number fledged by the trio. Analysing differences in seasonal reproductive success, polyandrous trios where two males fed the brood produced significantly more young per season than trios where only the alpha male fed[8] but the difference is clearly less marked over a season than per attempt.

These data on observed seasonal reproductive success are still unsatisfactory, however, for two reasons. First, many birds changed mating combinations within a season (Chapter 4) and so sample sizes for individuals who remained in the same mating system over a whole season are rather small. For example, only eight females were involved in a polygynous mating system throughout a whole season (Table 9.1). Second, and most crucially, these changes in mating system are sometimes a consequence of particular behavioural strategies. If we want to understand the reproductive payoffs of behaviour we therefore need to take these changes into account.

The most important change to consider is the behaviour of beta males in relation to their mating success. During 1981−1984, there were 12 cases where the beta male in polyandry succeeded in copulating for the first breeding attempt of the season and in all 12 cases he remained on the territory for the second attempt. However, out of 19 cases where the beta male failed to copulate in the first attempt, he remained for a second attempt in only 13 cases and left to compete for a female in another territory in 6 cases.[9] The result of a beta male leaving is that the alpha male can then breed as a monogamous male for the rest of the season. Clearly, then, we need to take this change into account when we assess the seasonal reproductive success of alpha males who guard their females so closely that the beta male fails to mate.

9.4 Calculation of seasonal reproductive success

For the reasons discussed above, Alasdair Houston and I (mainly Alasdair!) used a model to calculate the reproductive success of individuals per season in the various mating combinations. Another advantage of using a model is that we can vary various parameters, such as the probability that a beta male leaves a trio if it fails to mate, to see exactly how different behaviour patterns adopted by individuals will influence their reproductive success.

We calculated reproductive success for a short season, where there was time for at most two successful breeding attempts, and for a long season, where there was time for a third attempt even after two successes. In the latter case the birds therefore have three successes if the last attempt is also successful. The sequences we used are listed in Table 9.3. We computed both a long and a short season because these gave values for seasonal reproductive success which spanned the observed values in Table 9.1.

The probability of occurrence of any given sequence in Table 9.3 is simply the product of the probability of each event in the sequence. In the simplest cases, where the mating combinations remained the same throughout the whole season, we regarded the probability of a success, s, as constant throughout the season and the probability, f, of a failure as $(1-s)$. Then, for example, the probability of the sequence success, failure, success, *SFS* is $s(1-s)s$. The sum of all the sequence probabilities in a season will equal one. These are given in Table 9.3. For each of the various mating combinations we used the data in the first column of Table 9.1 as a measure of s and the data in the second column for the number of young, n, raised per successful attempt. We could then calculate the probability of a given number of successes for each mating combination over the season and the expected number of young reared per season. For example, in the short season we sum the sequence probabilities in Table 9.3 to give the expected number of young reared as,

$$n \ (f^3s + 3f^2s) + 2n \ (2s^2f + s^2).$$

The procedure is a bit more complicated when changes occur in the mating combinations during the season. I shall discuss two kinds of change (for others, see Davies and Houston 1986).

(i) A beta male left a trio when it failed to mate, thus leaving the alpha male and female to breed as a monogamous pair for the remainder of the season. In

Table 9.3. Sequences of failed, F, and successful, S, breeding attempts used in the model to calculate reproductive success for both a short breeding season (only time for, at most, two successful breeding attempts) and a long breeding season (time for three successes). Below the list of possible sequences is given the probability of achieving zero, one, two or three successful breeding attempts in the season, with f = probability an attempt is a failure and s, a success (from Davies and Houston 1986).

	Short season	Long season
(a) *Sequences*		
	FFFF	FFFF
	FFFS	FFFS
	FFS	FFS
	FSF	FSF
	SFF	SFF
	FSS	FSS
	SFS	SFS
	SS	SSF
		SSS
(b) *Probability in season*		
No successes:	f^4	f^4
One success:	$f^3 s + 3f^2 s$	$f^3 s + 3f^2 s$
Two successes:	$2s^2 f + s^2$	$3s^2 f$
Three successes:	0	s^3

our first calculations (Table 9.4) we assumed the probability of the beta male leaving, L, to be 0.3, the average value observed (see above).

(ii) The value of s changes during the season, to represent, for example, changes in the interference probability by beta males (see Chapter 8).

For these cases, where we allowed for changes in mating combinations during the season, we constructed flow charts for all the possible sequences. We then calculated, as in Table 9.3, the probability of occurrence of each sequence and the number of young produced in each, and then summed these to give the overall reproductive success per season. For example, in case (i) above, after a beta male failed to mate it had a probability, L, of leaving. If it left, the alpha male and female then bred as a monogamous pair and the values of s and n changed accordingly. If the beta male stayed, then the values of s and n remained as for a trio where only the alpha male mated, until the next breeding attempt when there was once again a probability, L, that the beta male would leave.

9.5 Do individual reproductive payoffs make sense of behavioural conflict?

Table 9.4 gives the model's calculations of seasonal reproductive success for males and females in the various mating combinations. Do these payoffs make sense of the way males and females set up their breeding territories (Chapters 4 and 5) and play out their games of conflict at mating time (Chapter 6)? I shall first consider the simpler mating systems involving pairs and trios of either one male and two females or two males and one female. Polygynandry will be discussed afterwards.

Female behaviour and reproductive success

A female gains full maternity of each brood she rears (Chapter 7) so her reproductive success is simply the total number of young fledged from all her breeding attempts in a season. Females showed three clear behaviour patterns.

(a) *Aggression to other females*. A female has least success in polygyny, where she has to share the help of one male. Her aggression to the other female in the system therefore makes good sense; by driving her off or disrupting her

Table 9.4. Model's calculation of seasonal reproductive success (total number of young fledged) per female and per male for short and long breeding seasons in the various mating combinations (modified from Davies and Houston 1986). For paternity and maternity assumptions, see text.

Mating system	Number of young fledged per season			
	Short season		Long season	
	Per female	Per male	Per female	Per male
Polygyny ♂ 2♀	3.8	7.6	4.4	8.8
Monogamy ♂ ♀	5.0	5.0	5.9	5.9
Polyandry 2♂ ♀				
Only alpha male mates and feeds	4.4	α 4.4	4.9	α 4.9
		β 0		β 0
Both alpha and beta males mate and feed	6.7	α 3.7	8.9	α 4.9
		β 3.0		β 4.0
Polygynandry 2♂ 2♀	3.6	α 5.0	4.0	α 5.6
		β 2.2		β 2.4

breeding attempt she has an increased chance of claiming the male's full-time help and so enjoying the greater success of monogamy (Table 9.4).

(b) *Encouragement of copulations from beta males.* A female has greatest success of all when two males help her full-time in polyandry. Her encouragement of copulations by the beta male therefore also makes good sense. If the beta male fails to copulate, not only is success reduced because only one male helps to feed the brood, but success is also less than in monogamy because of interference costs. There is, therefore, a double incentive for the female to get the beta male to mate. Her active attempts to avoid the alpha male's exclusive guarding is exactly what we would expect from the female's reproductive payoffs in Table 9.4.

(c) *Females ignored males when first setting up their breeding territories.* There was no indication in Chapter 4 that females were intent on gaining two males in the first place. They appeared to disregard male distribution altogether when setting up their territory boundaries with other females. Furthermore, they responded to the feeder experiment by decreasing their territory size, which made it less likely that they would associate with two males. Why does a female not attempt to defend as large a territory as possible, to increase her chance of gaining a second male, rather than simply setting up a territory in relation to resources and then leaving it to the males to sort out how many will associate with her?

The data on reproductive success suggest an answer. Although females can indeed gain from polyandry if both males mate, they suffer compared to monogamy if only the alpha male mates (Table 9.4). Overall, the average reproductive output from a nest in polyandry (including cases where one and two males mate/feed the brood) is a success rate of 64% and an average of 2.09 young per breeding attempt, no different from monogamy.[10] There is thus no incentive for a female to gain two males in the first place. Once she has two males, however, it certainly pays her to try and gain copulations from both so as to induce both to care.

In summary, the reproductive payoffs in Table 9.4 make excellent sense of all three female behaviour patterns.

Male behaviour and reproductive success

Where one male defends a territory he is likely to have full paternity of the broods raised (Chapter 7) so I have assumed the resident male is father of all the chicks in monogamy and polygyny. In polyandry, I have assumed that the alpha male gains full paternity in cases where only he mated and fed the brood, and that the paternity share was 55% alpha:45% beta in cases where both males mated (as shown by the DNA fingerprinting results in Chapter 7, Section 7.4). Males showed four distinct behaviour patterns.

(a) *Attempts to gain a second female*. Table 9.4 shows that a male does best with polygyny, the system in which a female does worst! Although, because of less help, each female raises fewer young, the total production of two polygynous females exceeds that of one monogamous female. A male's attempts to monopolise a second female (Chapters 4 and 5) therefore pays, as do his attempts to prevent disputes between his two females.

(b) *Attempts to drive off beta males*. Monogamy is more profitable than the alpha male's success in polyandry where he prevents the beta male from mating, because of interference costs. Monogamy is also more profitable than shared paternity in polyandry because, although more young are raised in total when two males cooperate to feed the brood, the increased production is not sufficient to offset the alpha male's paternity loss. The attempts to drive off beta males therefore is what we would predict.

(c) *Beta males leave polyandry to gain monogamy*. From the beta male's point of view, it is clearly better to copulate in polyandry but the payoff from shared paternity is much less than that in monogamy. As predicted, whenever a neighbouring female becomes available through the death of her mate, a beta male is always quick to leave a polyandrous trio to try and claim her (Chapter 4).

(d) *Alpha males in polyandry attempt to gain full paternity*. If an alpha male cannot drive a beta male away altogether, to claim the increased success from monogamy, should he ever agree to share paternity to avoid interference costs? There was no indication whatsoever that alpha males ever deliberately allowed beta males to copulate (Chapter 6). Shared matings, and the resulting co-operation between alpha and beta males in chick feeding, simply arose by default whenever alpha males failed to prevent the beta male from gaining access to the female.

Table 9.4 shows that the alpha male's selfishness makes good sense from the calculations for a short season because his success is greater in polyandry if he monopolises all the matings rather than sharing paternity. Thus it is better for him to sustain interference costs rather than costs of shared paternity. For the long season, the payoffs from the selfish strategy are the same as for cooperation. These figures are based on an average failure probability, f, of 0.58 throughout the season for polyandry where only the alpha male mates (Table 9.1). Incorporating the seasonal decline in f, which reflects higher interference earlier on (Table 8.6) makes little difference to the calculations.

These values for reproductive success of polyandry where only the alpha male mates are based on the observed average probability of 0.3 that a beta male leaves the trio if he fails to mate. As this value increases, there is an increase in the critical paternity which an alpha male must achieve for cooperation to pay. For example, with $L = 0.5$ an alpha male would have to gain more than 70%

A beta male, about to copulate, is interrupted during cloaca pecking by the approaching alpha male (behind), who will chase him away from the female's vicinity. It pays the female to mate with the beta male to gain his help with parental care. However, calculations show that the increased production of a trio-fed brood does not compensate the alpha male for the cost of shared paternity. As predicted, alpha males try to monopolise all the matings if they can.

of the paternity in a short season to make cooperation better than selfish exclusive guarding (Davies and Houston 1986).

The way paternity is shared is thus critical for deciding whether it is best for the alpha male to be selfish or cooperate in polyandry. Depending on the values used for various parameters in the model (season length, probability failed beta males leave, failure rate), the critical paternity the alpha male needs for cooperation to pay is 53–70% of the production of a trio-fed brood. With an average observed paternity share of only 55% it seems as if under most circumstances this critical threshold is not reached.

Certainly, if the alpha male could claim more than 75% of the paternity, then he would do better with cooperative polyandry than with monogamy, assuming the same number of young are raised as in Table 9.4. Under these circumstances it would pay the alpha male to agree to share the female. However, it is unlikely that the alpha male could ever impose a strict limit on the beta male's access. The alpha male could not simply sit on a perch, watch the beta male copulate, and then chase him off once he has performed, say, 25% of the matings. Obviously, it would pay the beta male to go for a larger share than the alpha male wanted to give him. Furthermore, from the female's point of view a 75:25% paternity split in favour of the alpha male may not maximise her reproductive success. I shall show in the following chapters that in polyandry each male works harder the greater his paternity share and that a 50:50 split maximises the total work males put into the brood. This is obviously the best share from a female's point of view.

In conclusion, the conflicts between alpha and beta male are what we would predict from their reproductive payoffs in Table 9.4.

9.6 Polygynandry: behaviour and reproductive success

The way males share matings and chick feeding in polygynandry is very variable and merits a whole chapter to itself (Chapter 11). I have included the model's calculations of reproductive success in Table 9.4 for the sake of completeness, but these average values obscure the interesting variation within this system and probably have little meaning because they are unlikely to represent the payoffs facing particular individuals.

On average, females do no better in polygynandry than in polygyny, which is surprising given the fact that they have two males rather than one. The high failure rate is primarily responsible for their low success and reflects a female's attempts to gain increased male care by interference (Chapter 8).

For males, I have assumed that alpha males gain on average 70% of the paternity, as shown in Chapter 7 (see Table 7.2). Three male behaviour patterns result in the formation of polygynandry (Chapter 4). Do the average values for male reproductive success in Table 9.4 make sense of these behavioural choices by individuals?

(a) *Beta male joins polygyny*. The resident male suffers because the payoffs for an alpha male in polygynandry are less than that from polygyny. The observed attempts by the resident male to resist beta male settlement therefore make good sense.

(b) *Polyandrous males expand to gain a second female*. On average, half the cases of polyandry involve the alpha male monopolising mating access and half involve shared mating access and cooperative brood care (Table 9.1). If we assume that male reproductive success is the average of the two cases of polyandry in Table 9.4, then both alpha and beta males gain from a change to polygynandry.

(c) *Monogamous male expands to take-over neighbouring female.* The values in Table 9.4 suggest that, on average, polygynandry is no better for an alpha male than is monogamy. However, in this case the average payoff for poly-gynandry certainly does not represent the individual gains involved. As shown in Chapter 4, monogamous males usually expand to takeover a neighbouring female only when their own female is incubating. Their payoff from expansion is thus added to that already secured from one case of full paternity.

9.7 Sexual conflict and the variable mating system

In general, the individual reproductive payoffs from the various mating systems make beautiful sense of the behavioural preferences shown by individuals and illuminate the reasons for the resulting conflicts of interest. At one level we may not be surprised at the exquisite fit between behaviour and reproductive success; this is exactly what we would predict from natural selection. However, the aim of measuring reproductive success was not simply to see whether selection has done a good job on moulding dunnock behaviour in a way which maximises individual success. The main aim was to understand why there were such intense conflicts of interest. Only by measuring offspring production from the different mating systems, together with maternity and paternity, was it possible to identify the sources of the conflict. The two main sources are: (1) The increased produc-tion of young in cooperative polyandry does not compensate a male for the fact that paternity is shared, hence monogamy is more profitable for males than cooperative polyandry, even though the reverse is true for females; and (2) despite the cost each polygynous female suffers from having to share a male's parental care, the combined output of two females exceeds that of one, hence polygyny is more profitable for males than monogamy, even though the reverse is true for females.

In discovering these sources of the conflicts of interest we can now also suggest an explanation for why the dunnock has such a variable mating system. The different systems may simply reflect the different outcomes of conflicts of interest (Fig. 9.2). Where a female is able to gain her best option, despite the conflicting interests of the alpha male, we observe cooperative polyandry with two males sharing matings and parental care. Where a male is able to gain his best option, despite the conflicting interests with each of his females, we observe polygyny. Monogamy is a mating system where neither sex has been able to gain a second mate. Polygynandry, for example two males with two females, can be viewed as a kind of 'stalemate'; the alpha male is unable to drive the beta male

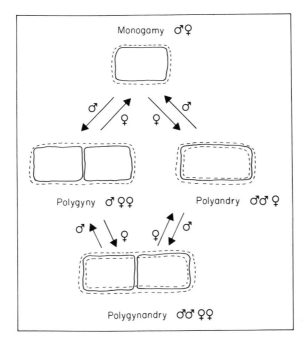

Fig. 9.2 Diagrams to show how male (– – –) and female (———) territories overlap in the various mating combinations, with arrows indicating the direction in which alpha male or female behaviour encourages changes in the mating system. From Davies 1989. Different mating systems emerge depending on which individuals achieve their preferred options.

off and so claim both females for himself, and neither female is able to evict the other and so claim both males for herself.

The interesting question to ask is, 'under what circumstances will particular individuals gain their preferred options despite the conflicting preferences of others?' The previous chapters have shown how ecological conditions and individual competitive ability influence the outcomes of the conflict. A female is more likely to achieve her best option when the sex ratio is male-biased, when the vegetation on her territory is dense (so she can escape the alpha male's close guarding) and where her two males are of more equal competitive ability (so both can gain matings). A male is more likely to achieve his best option if he can defend a larger territory and when he is older and so more familiar with his territory and competitors. The main point is that, with conflicting interests, not all individuals will achieve their preferred option and so we need to consider the options available to each party. I shall discuss the general relevance of sexual conflict in the final chapter.

9.8 Summary

Short-term measures of reproductive success give a good indication of an individual dunnock's fitness gain from following different courses of action because there was no difference in adult survival with mating system. Success per season is a better short-term measure than success per breeding attempt because it compares the number of young produced over a given time period.

Individual reproductive payoffs from the different mating systems make excellent sense of the behavioural preferences shown by males and females and also illuminate the sources of the conflicts of interest described in the previous chapters. For females, reproductive success is greatest with cooperative polyandry, less with monogamy and least with polygyny. A female's aggression towards other females and her encouragement of beta male copulations therefore both make good sense.

For males, however, reproductive success is greatest with polygyny, less with monogamy and least with cooperative polyandry (the reverse order of success for females). A male's attempts to gain additional females and to prevent other males from sharing matings, therefore, make good sense.

It is suggested that the variable mating system of the dunnock reflects the different outcomes of these conflicts of interest among males and females.

STATISTICAL ANALYSIS

1. Considering broods fledged in April, May, June and July (n = 250), χ^2_3 = 2.687, $P > 0.30$.
2. Considering young fledged in April, May, June and July (n = 723), χ^2_3 = 4.596, $P > 0.20$. There was no significant variation between years in seasonal success.
3. Kruskal-Wallis 1-way ANOVA, H_5 = 0.681, $P > 0.90$.
4. Kruskal-Wallis 1-way ANOVA, H_4 = 4.394, $P > 0.50$.
5. Correlation between date present successful attempt began and time to start of next attempt, r = −0.041, n = 60, NS.
6. t-test, t_{103} = 7.521, $P < 0.001$, 2-tailed.
7. Predation, 29.0 days (SD 8.5, n = 28). Interference, 26.9 (SD 7.2, n = 16) days. t_{43} = 0.847, NS.
8. Mann-Whitney U-test, P = 0.023, 2-tailed.
9. Beta males more likely to leave if they fail to copulate; Fisher exact test, $P < 0.05$.
10. Comparing overall success rate of polyandry (one and two males mated combined) and monogamy in Table 8.4 (previous chapter), χ^2_1 = 0.101, $P > 0.50$. Comparing mean number of young fledged per attempt, P = 0.452, Mann Whitney U-test, 2-tailed.

10

Parental effort by males and females in pairs and trios

10.1 The evolution of stable cooperation

We ended Chapter 9 with the pleasing conclusion that the observed conflicts of interest between individuals make good adaptive sense, given their reproductive payoffs from the various mating systems. It might be thought that we could now end the book with the claim that the dunnock puzzle has been solved. However, all we have really done is to push the problem back a step. Why have the reproductive payoffs ended up at the particular values we observe? Why, for example, does the beta male not help more with chick feeding so that it pays the alpha male to cooperate? If this happened, then there would be no male-female conflict over the occurrence of cooperative polyandry. Likewise, why does a polygynous male not help his females more so that they do not suffer costs from reduced chick care? If this happened, then there would be no male-female conflict over the occurrence of polygyny.

To answer questions like these we need to think in more general terms about the evolution of life histories, in particular the trade-offs between survival and reproductive success. How hard should an individual work at feeding its young? The harder it works the more likely the young are to survive, but harder work is likely to decrease the adult's own chances of survival. In many cases, some intermediate level of effort would therefore seem to be best (Williams 1966*b*; Nur 1984). When only one parent provides care, as in many fish and mammals, the optimal level of parental investment can be calculated from a knowledge of the trade-off between the gains in fitness from the current brood and the costs of investment for the parent's future survival and reproduction.

However, where two parents provide care, as in most birds, the optimum effort for one individual will also depend on how hard the other is prepared to work. Recent observations on several species of birds have shown that either member of the pair is capable of increasing its provisioning rate to the young if the other partner deserts or is removed experimentally (e.g., great tits *Parus major* and blue tits *Parus caeruleus*, Sasvari 1986; snow buntings *Plectrophenax nivalis*, Lyon *et al.* 1987; pied flycatchers *Ficedula hypoleuca*, Alatalo *et al.* 1988*b*; starlings *Sturnus vulgaris* Wright and Cuthill 1989). Similar reactions may occur where more than two adults care for a brood. For example, in cooperative breeders, such as grey-crowned babblers *Pomatostomus temporalis* (Brown *et al.* 1978) and moorhens *Gallinula chloropus* (Gibbons 1987), the parents invest less in brood care if they have helpers (previous offspring) contributing to chick provisioning.

These observations raise a problem. If individuals will work harder when others do less, and less hard when others do more, how do two or more individuals come to a stable agreement on how much work each should do? The problem is that of cheating; each individual may be tempted to get away with less than its fair share of work and rely on the compensatory reactions by others.

Alasdair Houston and I have used a model to help us think about this problem (Houston and Davies 1985), developing an idea first suggested by Chase (1980). Consider a pair of birds. The male will have a 'best response', for example in terms of chick feeding effort, to a given effort put in by the female. If the female works harder, it will pay the male to do less and vice versa (Fig. 10.1a). The female will likewise have a 'best response' to what the male is prepared to do (Fig. 10.1b). In both cases, the reaction lines have been drawn with a slope of less than −1, which means that if one parent does less work, the other responds by doing more, but the reaction is not sufficient to compensate fully for the loss. This is the pattern seen in the studies mentioned above; when one partner is removed, the other increases its provisioning rate to the young but the total effort of a single parent is less than that of a pair.

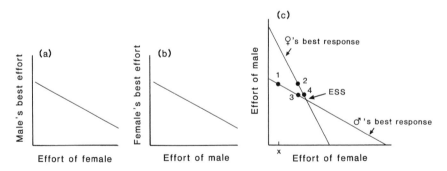

Fig. 10.1 Stable cooperation in chick feeding. From Houston and Davies 1985. (a) It pays the male to work harder the less the female does. (b) It pays the female to work harder the less the male does. For simplicity, these 'reactions' have been drawn as straight lines. (c) The male and female reaction lines plotted together. Provided the slopes of the lines are less than 1, the intersection point is the stable solution to the conflict. ESS = evolutionarily stable strategy. See text.

Given that each member of the pair has a reaction curve, what will be the result? To see this, we simply plot both responses on the same graph (Fig. 10.1c). The intersection point turns out to be the stable solution to the conflict. For example, if the female played effort x, then the male's best response would be 1. The female would then reply with 2, the male with 3 and so on, reactions proceeding by smaller and smaller amounts until the intersection point was reached. This is the stable effort for male and female because at this point it does not then pay either sex to change. The same kind of analysis can be done for three cooperating individuals, to represent cooperative polyandry in the dunnock. In this case the optimum effort for one adult is a function of the efforts of the other two. Once again a stable equilibrium can be reached, with all three adults caring, provided the reactions are not completely compensatory. In other words, lazy partners would be penalised because their reduction in effort would lead to a lower total effort and hence poorer chick nourishment and survival.

Other outcomes are possible depending on how the reaction curves meet (see also Winkler 1987). For example, if the curves have a slope of greater than -1, the intersection point is unstable. Here, if one parent reduces its effort, the other reacts by increasing its effort to more than compensate. The first parent will then be tempted to reduce its effort further and reactions will proceed by larger and larger amounts until one parent ends up doing all the work.

In theory, the stable solution to the conflict could be reached by males and females reacting to each other's bids on a short time-scale. Some of the complex courtship seen between breeding adults could, perhaps, reflect this kind of bargaining. Alternatively, we may not actually see any conflict now because the game may have already been played over evolutionary time, with individuals now designed by natural selection to play the stable effort.

The aim of the next three chapters is to consider the factors which influence parental effort. In this chapter I shall present data on the provisioning efforts of individuals when working in teams of two (pairs) and teams of three (trios). I shall use provisioning rate as a measure of parental effort because, as shown in Chapter 8, this provides a good measure of benefits to nestlings and is also likely to reflect costs to parents (Nur 1984; Bryant 1988). I shall then consider how individuals react to the efforts of others to test whether, as predicted by the model, stable cooperation in pairs and trios arises because reaction responses are insufficient to fully compensate for reduced investment by others.

10.2 Provisioning of nestlings by pairs and trios

Provisioning rates were measured by watching nests for on average 2.7 hours each when the chicks were 5−11 days of age, when provisioning was at its peak rate. Pair-fed broods include cases where one male and one female fed full-time at a nest in monogamy, and also cases of polyandry and polygynandry where the female was helped full-time by only one male. There were no differences

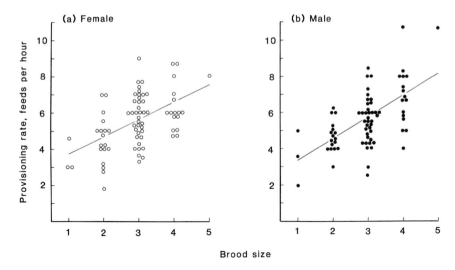

Fig. 10.2 Provisioning rate to the brood in relation to brood size for 75 pair-fed broods of nestlings. (a) Female provisioning rate; $y = 2.86 + 0.92x$; $r = 0.493$, $P < 0.01$. (b) Male provisioning rate; $y = 2.15 + 1.19x$; $r = 0.594$, $P < 0.01$. No difference between male and female rates (ANCOVA; $F_{1,146}$ slope $= 1.03$, NS; $F_{1,147}$ elevation $= 0.15$, NS).

in female or male provisioning rates across these three cases (Hatchwell and Davies 1990) and they have been combined in Fig. 10.2. This shows that, in pairs, males and females provisioned at the same rate, both increasing their effort in the same way as brood size increased.

Trio-fed broods include cases where two males helped a female full-time in polyandry and polygynandry. There were no differences between these two (Hatchwell and Davies 1990) and they have been combined in Fig. 10.3. All three adults increased their provisioning rates with brood size, with no differences in their rates of increase. Alpha males and females provisioned at the same rate, and their rates were no different from those of paired males and females respectively.[1] Beta males provisioned at a significantly lower rate than the alpha male and female. As shown earlier (Chapter 7), beta males tend to have lower paternity than alpha males and I shall demonstrate experimentally in Chapter 12 that their lower provisioning rate is caused by their lower assessment of paternity. However, it is interesting to note that beta males do not drop out at lower brood sizes, showing the same slope of reduction as the other two adults, so that all three continued to provision even a single nestling.

The greater provisioning rate in trios compared to pairs (Chapter 8, see also Fig. 10.6) arises, therefore, simply because the contribution of the beta male

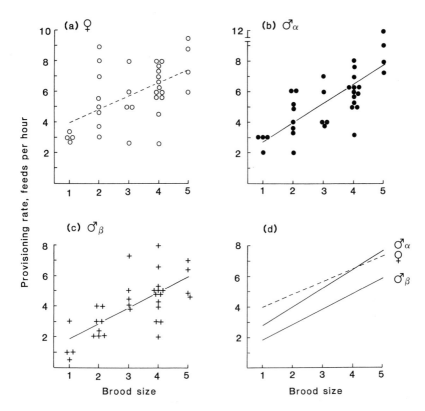

Fig. 10.3 Provisioning rate to the brood in relation to brood size for 34 trio-fed broods of nestlings. (a) Female provisioning rate; $y = 3.09 + 0.86x$; $r = 0.542$, $P < 0.01$. (b) Alpha male provisioning rate; $y = 1.48 + 1.25x$; $r = 0.749$, $P < 0.01$. (c) Beta male provisioning rate; $y = 0.79 + 1.03x$; $r = 0.701$, $P < 0.01$. (d) The three regression lines for comparison. They differ significantly in elevation $F_{2,98} = 13.54$, $P < 0.01$, but not in slope $F_{2,96} = 0.90$, NS. No difference between the female and alpha male lines (Tukey's test, $q = 1.48$, 96df, NS) but beta male rates significantly lower than both that of the female ($q = 6.97$, 96df, $P < 0.01$) and alpha male ($q = 5.49$, 96df, $P < 0.01$).

adds on to that of the alpha male and female, who are each provisioning at the same rate as in a pair (see also Byle 1991).

10.3 Reactions to changes in effort by others

The neatest test of how individuals respond to reduced effort by their partners is the work of Wright and Cuthill (1989), who manipulated work rates of starlings by adding small weights to the base of the tail. For the dunnocks, Ben

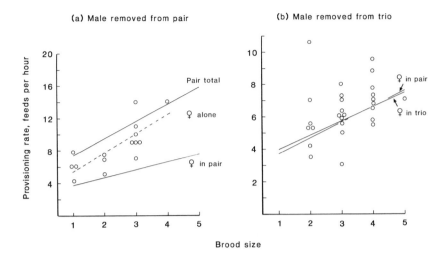

Fig. 10.4 Female reactions to changes in male effort. (a) When a male was removed from a pair ($n = 15$), the female's provisioning rate ($\circ$; $y = 2.96 + 2.33x$; $r = 0.783$, $P < 0.01$) increased above that of a female in a pair (lower line; $F_{1,86}$ slope $= 10.43$, $P < 0.01$; $F_{1,87}$ elevation $= 67.13$, $P < 0.01$). However, despite the increase, it was still lower than the total rate of a pair ($F_{1,86}$ slope $= 0.15$, NS; $F_{1,87}$ elevation $= 6.56$, $P < 0.05$). (b) When a male was removed from a trio ($n = 27$), the female's provisioning rate ($\circ$) did not differ significantly from the rates of pair or trio females (sign tests, NS). From Hatchwell and Davies 1990.

Hatchwell and I investigated only larger scale responses to the removal of a partner, mainly caused by death, predation or desertion, supplemented by some temporary experimental removals (Hatchwell and Davies 1990).

When paired females were left alone, they increased their effort but it was not sufficient to compensate fully for the loss of the male so the total rate was still less than that of a pair (Fig. 10.4a). As a result, chick weight was reduced compared to pair-fed broods (see Fig. 8.3). By contrast, when one male was removed from a trio, females did not react but maintained the same rate (Fig. 10.4b). This result might have been anticipated, given the lack of difference in provisioning rates by females in pairs and trios, found above.

When a beta male was removed from a trio, we could not detect any reaction by the alpha male (Fig. 10.5a). Again, this might have been expected from the lack of difference between alpha male and paired male rates. However, when an alpha male was removed from a trio, the beta male increased his effort to the same level as that of a paired male (Fig. 10.5b). Following removal of one of the trio males, reactions (when they occurred) were not sufficient to compensate for the loss so that the total effort was still less than that of an intact

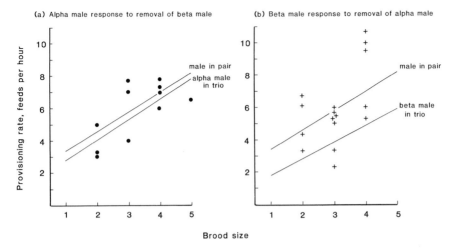

Fig. 10.5 Male reactions to changes in male effort resulting from the removal of a male from a trio. (a) When a beta male was removed ($n = 11$) the alpha male's provisioning rate ($\bullet$; $y = 1.96 + 1.20x$; $r = 0.684$, $P < 0.05$) did not differ significantly from that of paired males ($F_{1,82}$ slope $= 0.00$, NS; $F_{1,83}$ elevation $= 0.15$, NS) or alpha males in trios ($F_{1,41}$ slope $= 0.02$, NS; $F_{1,42}$ elevation $= 0.36$, NS). (b) When an alpha male was removed ($n = 16$), the beta male's provisioning rate ($+$; $y = 0.71 + 1.71x$; $r = 0.553$, $P < 0.05$) increased above that of a beta male in a trio ($F_{1,46}$ slope $= 1.41$, NS; $F_{1,47}$ elevation $= 17.37$, $P < 0.002$) to a level no different from that of a pair male ($F_{1,87}$ slope $= 0.97$, NS; $F_{1,88}$ elevation $= 0.12$, NS). From Hatchwell and Davies 1990.

trio (Fig. 10.6). As a result, chick weight was reduced compared to trio-fed broods (see Fig. 8.3).

The nature of these responses cannot be understood without knowing exactly how provisioning rate influences both nestling and adult survival. Nevertheless, three conclusions can be made.

(a) The occurrence of reactions seems related to the needs of the young. Presumably the relationship between provisioning rate and nestling survival in curvilinear, reaching an asymptote as increased provisioning saturates the requirements of the nestlings and so brings no further benefit. Some evidence for this is provided by the fact that there is a steep increase in survival to independence at low nestling weights, but this begins to level off with further weight increase (see Fig. 8.4). This may explain why females increase their provisioning only when male help is reduced below the level of one male help. The cost to chicks of a drop from two male help to one male help is much less than that from one male help to no male help, and compensation might not pay in the former case if this increases risk to the female herself.

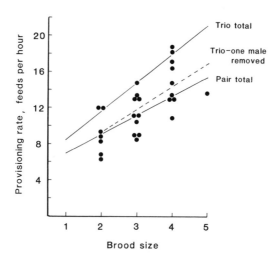

Fig. 10.6 When one male was removed from a trio ($n = 27$, • and $---$; $y = 3.96 + 2.58x$; $r = 0.684$, $P < 0.001$), the total rate, despite any compensation reactions, remained lower than that of an intact trio (top line, $F_{1,57}$ slope $= 0.58$, NS; $F_{1,58}$ elevation $= 19.00$, $P < 0.002$) and was no different from that of a pair (bottom line, $F_{1,98}$ slope $= 0.71$, NS; $F_{1,99}$ elevation $= 0.81$, NS). From Hatchwell and Davies 1990.

This may also explain why alpha males do not compensate when beta males are removed but beta males do so when alpha males are removed. The loss of a beta male's contribution is less and may not merit a reaction if this is costly to the remaining male.

(b) The fact that reactions occur will influence the profitability of desertion. Thus, the increase in effort by lone females will make it more profitable for a monogamous male to desert and gain a second mate (polygyny). Likewise, the compensating reaction by the beta male will make it more profitable for a polyandrous male to desert and monopolise a second female, thus changing the system to polygynandry. In both cases the remaining adults seem to be forced into a 'cruel bind' (Trivers 1972). It is particularly interesting that beta males are prepared to increase their effort to the level of a monogamous male, despite the fact that they have only part paternity of the brood (see Chapter 12 for further discussion).

(c) As in other studies, the magnitude of the reactions was not sufficient to compensate fully for the loss of a partner. The model in Fig. 10.1 showed how such incomplete reactions could lead to cooperation as the stable outcome, with both members of the pair provisioning in monogamy and all three adults provisioning in polyandry, as observed.

A kestrel Falco tinnunculus, *one of the predators of small birds in the Garden. When an adult is killed by a predator or removed temporarily by experiment during chick feeding, other adults attending the brood increase their parental effort but not sufficiently to compensate fully for the lost help.*

10.4 Care of fledglings: brood division

Once the young left the nest, they continued to be fed for another 2 to 3 weeks before they became independent. Fledglings were difficult to observe because they hid away in dense vegetation and it was not possible to get the same kind of detailed data as obtained by David Harper (1985) in his brilliant study of fledgling care by robins *Erithacus rubecula* in the Garden.

Of 18 broods of fledglings studied by Philip Byle (1990), 6 were 'undivided', with all chicks fed by all attendant adults, while 12 were 'divided', with each parent taking major or sole charge of part of the brood for most or all of the fledgling period. When the brood was divided, each adult tended to go off with their chick(s) to a different part of the territory and the feeding ranges of the adults overlapped much less than when broods were undivided. There was no difference in brood size between divided and undivided broods, nor in nestling weight (day 6), but like Harper's robins, broods were more likely to be divided later in the season, when there were no further nesting attempts to follow. In robins the occurrence of brood division seems partly related to food abundance because it broke down when extra food was provided, and the greater frequency of divided broods later in the season may have been related to lower food availability (Harper 1985). Even so, it is not clear why food abundance should influence brood division. In the dunnocks the provision of food had no clear effect on brood division and the provisioning rate of fledglings in divided and undivided broods did not differ (Byle 1990).

Including subsequent observations together with those obtained by Byle, there were 13 cases where one male and one female attended a divided brood. In 8 cases there were two fledglings, with the parents looking after one each. In 4 cases there were three fledglings; in 2 of these the female looked after two young and the male one, while in the other 2 cases the male looked after two and the female one. In one case there were four fledglings and each parent looked after two young. Thus, as in nestling care (Fig. 10.2), males and females divided the parental effort equally (mean percentage of fledglings cared for by the male = 50%, SE 2.8).

In trio-fed broods there was also evidence that parental effort for fledglings followed the pattern set at the nestling stage, with beta males putting in less effort than the alpha male and female. Of 13 broods studied by Byle (1990), where a trio fed the nestlings, in only 7 cases did the beta males continue to help at the fledgling stage. They were more likely to do so where there was a larger number of fledglings. Including subsequent data, there were 7 cases of broods being divided among three adults in polyandry or polygynandry. In 2 cases there were three young, with the adults taking one young each. In 5 cases there were four young; in all 5 the beta male attended just one young with the female taking two and the alpha male one in 2 cases and the alpha male taking two and the female one in 3 cases. Although on this small sample size the division of the brood among the three adults was not significantly biased,[2] there was a tendency for beta males to do less work.

10.5 Summary

In pairs, there was no difference between male and female provisioning rates to broods of nestlings. Where two males and a female fed the brood, the alpha male and female provisioned at the same rate as a male and female of a monogamous pair, and the beta male provisioned at a lower rate.

Females did not change their provisioning rate when male help was reduced from two males to one male, but they increased when help was further reduced to no male help. When a beta male was removed from a trio, the alpha male did not change his provisioning rate. However, when an alpha male was removed the beta male increased his rate to that of a monogamous male.

These reactions show that individuals will sometimes work harder if others do less and raise the question of whether it would pay to cheat by reducing effort and so relying on compensation responses by others. However, when reactions occurred they were not sufficient to compensate fully for the loss of the removed adult and nestling weight was reduced. A model shows how such 'incomplete' compensation can lead to cooperation as the stable outcome, with both members of a pair provisioning in monogamy and all three adults provisioning in polyandry, as observed.

Fledglings were sometimes divided among the parents for care to independence. Fledgling care followed the same pattern as nestling care, with males and females dividing the brood equally in pairs and beta males doing less work than alpha males and females in trios.

STATISTICAL ANALYSIS

1. Provisioning rates of alpha males in trios (Fig. 10.3b) were no different from those of males in pairs (Fig. 10.2b); ANCOVA, $F_{1,105}$ slope = 0.07. NS; $F_{1,106}$ elevation = 2.81, NS. Provisioning rates of females in trios (Fig. 10.3a) were no different from those of females in pairs (Fig. 10.2a); ANCOVA, $F_{1,105}$ slope = 0.04, NS; $F_{1,106}$ elevation = 0.04, NS. (See Hatchwell and Davies 1990).

2. For the 7 broods divided among the three adults, the mean percentage of the brood tended by each adult was: female 35%, alpha male 38%, beta male 27%. Friedman 2-way ANOVA, $\chi^2_2 = 1.50$, $P > 0.30$.

11

How males allocate effort between broods in polygyny and polygynandry

11.1 The choices facing males

In Chapter 9 I considered the question of which mating system best maximised an individual's reproductive success and concluded that for males it was best to breed with more than one female. This chapter, like the previous one, attempts to go back a step and ask, given the choice of mating system how should a male best allocate his mating and parental effort? It is the outcome of these choices which determines the reproductive payoffs and so sets the stage for the behavioural conflicts described earlier in the book.

Depending on the reproductive synchrony of the females in polygyny and polygynandry, the males may have to choose between two egg laying females during the mating period (trade-off between two mating efforts), between two synchronous broods of young (trade-off between two parental efforts) or between an egg laying female and a brood of young (trade-off between mating and parental effort). To test whether males allocate effort between these options in a way which maximises their reproductive success, I first quantify the value of male parental care in terms of how it influences reproductive output. This then makes it possible to predict how a male should best choose between the various alternative options above.

11.2 The value of male parental care

The amount of male help with chick feeding varied from none at all, through one male part-time, one male full-time, two males part-time and two males full-time, depending on the mating system (see Section 8.2). A male's help was classified as 'part-time' if, during any part of the nestling period he left to associate with another female or brood. Likewise, when two males helped, if either or both deserted for part of the nestling period their help was scored as 'part-time'. Within these five categories of amount of male help, there was no significant variation between mating systems in either provisioning rate or number of young fledged per brood (Davies and Hatchwell 1992). Therefore, data from the different mating systems were combined for each of the five categories of male care. Table 11.1 shows that there was no significant variation in the number of young hatched with amount of male help, but highly significant variation in the number of young fledged, with increased success when there was more male help. This was not due to any difference in predation but rather to reduced nestling starvation as male help increased. As shown by removal experiments in Chapter 8, these differences in fledging success were directly caused by differences in amount of male help.

Figure 11.1 summarises the effects of male help in terms of the number of young fledging successfully in relation to the number of young hatching. This makes two points. First, male help becomes increasingly valuable as brood size increases. This can be seen from the fact that the slopes of the lines increase with increasing amounts of male help, which mean that the effects of male help become more marked at larger brood sizes. A female on her own can raise one hatchling but she needs help to have a chance of fledging more than one, and the more help the better, especially with larger broods. Second, for all brood sizes the addition of one male's help is more valuable than the addition of a second male's help. As we shall see, these two conclusions have important implications for how a male should best allocate his effort between broods.

11.3 Reproductive allocation by males in polygyny

It was unusual for one male to gain exclusive access to two females, so there are far fewer data on polygyny than polygynandry (Chapter 3, Table 3.2).

In all 3 cases of polygyny where two females were simultaneously in their mating period, a male spent time with both. In all 5 cases of polygyny where one female had young and the other was offering matings, the male deserted the young and spent all his time with the mating female. Thus polygynous males preferred matings over chick feeding. The payoffs in Table 11.1 indicate that this makes good sense. If a male can claim full paternity of two broods he does better in polygyny than in monogamy provided he either helps one brood full-time (in which case he increases his success over monogamy by the 0.81 young reared by his other female whom he abandons), or helps both females part-time (where he gains $2 \times 1.63 = 3.26$, which is greater than the 2.41 from monogamy. It clearly makes no sense for a male to gain two females but to then help neither (payoff $= 0.81 \times 2 = 1.62$, less than monogamy). As predicted, polygynous males always helped with parental care.

How should a polygynous male allocate parental effort between two synchronous broods? Table 11.1 suggests that a male will on average do equally well helping one female full-time and leaving the other female to care alone

Table 11.1. Variation in reproductive success per nest with amount of male help during feeding of nestlings (from Davies and Hatchwell 1992).

Measures of reproductive success	Amount of male help					Difference across the 5 categories
	No male help	One male part-time	One male full-time	Two males part-time	Two males full-time	
No. nests	31	27	153	31	57	
Mean no. young hatched (SE)	2.89 (0.20)	3.04 (0.17)	3.24 (0.07)	3.17 (0.16)	3.46 (0.13)	$F_{4,281} = 2.060$ NS
% nests depredated	16.1	14.8	16.9	19.3	12.3	$\chi^2_4 = 0.978$ NS
% nests where some/all young starved to death	80.8	73.9	26.8	40.0	6.0	$\chi^2_4 = 62.61$ $P < 0.001$
Mean no. young fledged (SE)	0.81 (0.17)	1.63 (0.22)	2.41 (0.12)	2.00 (0.24)	3.00 (0.19)	$F_{4,294} = 15.20$ $P < 0.001$

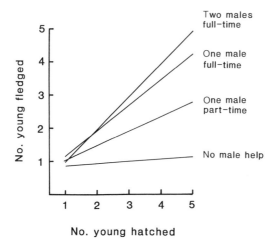

Fig. 11.1 Reproductive success (no. young fledged) in relation to number of young hatched for different degrees of male help with parental care. Regressions calculated from data in Table 11.1, excluding predation. In all cases linear regressions provided the best fits. Two males full-time, $y = -0.021 + 0.989x$. One male full-time, $y = 0.330 + 0.784x$. One male part-time, $y = 0.535 + 0.447x$. No male help, $y = 0.787 + 0.064x$. Two males part-time (not plotted), $y = 0.880 + 0.513x$. Comparing the 5 regressions there was significant variation in slope (ANCOVA; $F_{4,241} = 8.41$, $P < 0.001$) and elevation ($F_{4,245} = 32.80$; $P < 0.001$). From Davies and Hatchwell 1992.

(payoff $= 2.41 + 0.81 = 3.22$), as helping both females part-time (payoff $= 2 \times 1.63 = 3.26$). However, Table 11.2 shows that in all 7 cases the male preferred to feed one brood and left the other either unhelped (5 cases) or with little help (2 cases). In 5 cases he preferred the first brood to hatch. In the one case where the broods were equal in size he preferred the earlier brood. In 5 out of 6 cases where the broods differed in size he preferred to feed the larger brood.

As shown earlier in Fig. 11.1, the addition of one male's help to an unaided female had little value with small broods, but increased in value as brood size increased. Fig. 11.2(a) illustrates how the brood sizes at two nests affects whether it is better for a male to help full-time at one versus part-time at both. The prediction is that it is better to feed full-time at one nest unless brood sizes are equal, in which case it is (only marginally) better to feed part-time at both. In 6 cases the choice of brood sizes in Table 11.2 was such that the male was predicted to feed full-time at the nest with the larger brood; he did so in 5 cases and strongly preferred the larger brood in the other case. In one case (3 young in each brood) he was predicted to do very slightly better by part-time care

Table 11.2. The proportion of his feeds which a polygynous male brings to each brood for 7 cases of two synchronous broods. Nest 1 is the first brood to hatch (from Davies and Hatchwell 1992).

Case	Nest 1		Nest 2	
	No. young	% feeds	No. young	% feeds
(a)	3	71.9	3	28.1
(b)	2	82.1	1	17.9
(c)	2	0	4	100
(d)	3	100	4	0
(e)	1	0	4	100
(f)	5	100	1	0
(g)	2	100	1	0

at both and did so, though he strongly preferred one brood. In general, therefore, the tendency for a polygynous male to prefer one brood maximised his reproductive success.

Experimental studies have shown that polygynous males of other species are also more likely to assist the larger brood (Patterson *et al.* 1980; Whittingham 1989) or older brood (Lifjeld and Slagsvold 1990; Veiga 1990*b*). Brood size may also influence the probability of mate desertion in species with successive polygamy, with single parents being more likely to be left in attendance when brood size is small (Beissinger 1990).

11.4 Reproductive allocation by males in polygynandry

An example

The way in which alpha and beta males share matings and chick feeding in polygynandry is complex but fascinating. In my first season with the dunnocks I was completely baffled. In fact, some of the male interchanges between females were so sudden and unexpected that I first thought that I must have misread the colour-rings. For example, one day the alpha male would be with one female and the beta with the other; the next day they would suddenly swap mates! And chick feeding likewise seemed a complete muddle, with the two males sometimes both feeding one female's brood while at other times splitting up to each help just one female. It took most of the 10 years to reduce all this complexity to a few simple themes, so I believe we now finally understand what is going on.

It will be helpful to begin with an example. In Fig. 11.3 two males shared three females. Female$_1$ began breeding (nest A) well ahead of the other

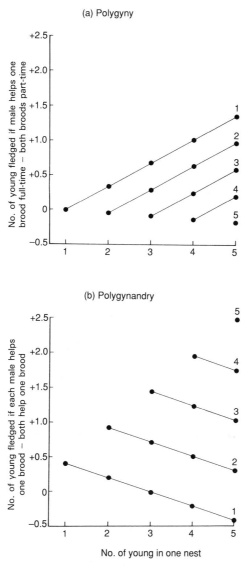

Fig. 11.2 Predicted differences in reproductive success arising from two ways of allocating parental care calculated from the linear regressions in Fig. 11.1. Various combinations of brood sizes are indicated, with number of young in one nest shown along the x-axis and number of young in the other nest represented by the various lines in each graph. (a) Polygynous males, who may either feed one brood full-time and leave the other brood unaided, or help both broods part-time. (b) Two polygynandrous males, who may either each help a different brood full-time or both help one brood full-time and leave the other brood unaided. From Davies and Hatchwell 1992.

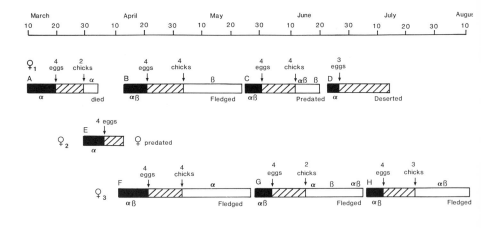

Fig. 11.3 An example of how two males (alpha and beta) in a polygynandrous mating system allocated their mating and parental effort among their three females during the breeding season. Each nesting attempt is indicated by a letter (A–H), which marks the stage where a female had a completed nest—the time at which matings began in earnest. The symbols α and β underneath each mating period (in black) and above each chick feeding period (in white) indicate which males gained matings and helped with chick feeding. See text for a detailed description of this saga.

two and the alpha male guarded her closely and monopolised all the matings. When the young hatched, only he helped the female to feed them. Meanwhile, female₂ had begun her first nesting attempt (nest E) and the alpha male then abandoned female₁'s chicks to guard this second female, where he likewise monopolised all the matings. During this time, and probably largely as a result of reduced help, all female₁'s chicks starved to death. Female₂ was then killed on the nest by a cat. So far, then, a month's hard work had produced nothing.

Female₁ then began a replacement nest B and, at the same time, female₃ began her first attempt of the season (nest F). Faced now with two synchronous females, the alpha male spent time with each and so the beta male simply joined whichever female was left unguarded. As a result, both males shared matings with both females. (In retrospect, my earlier observations of two males suddenly 'swapping' mates must have been cases like this, where an alpha male changed to guard another female in his system, thus allowing the beta male free access to the first.) The two broods hatched on the same day and the beta male helped female₁ while the alpha male helped female₃, with both broods fledging young successfully.

Both females then began another nesting attempt in synchrony, and once again the alpha male spent time with both thus allowing the beta male access to whichever he left unguarded at the time. Female$_1$'s chicks (nest C) hatched first, whereupon both males helped to feed them. A few days later, female$_3$'s chicks hatched (nest G) and the alpha male moved over to give her his full-time help, leaving the beta male to care for the chicks of female$_1$. Female$_1$'s brood was then predated, probably by a crow. When she started another attempt (nest D), the alpha male deserted female$_3$'s chicks and came over to guard her and monopolised all the matings, while the beta male took over the feeding of female$_3$'s brood. Once female$_1$ began incubation, the alpha male returned to help female$_3$, so she gained two male's help for the final period of chick rearing. Female$_3$'s brood was thus first fed exclusively by the alpha male, then exclusively by the beta male and finally by both males together.

To complete the saga, female$_1$ deserted nest D, her final attempt of the year. Female$_3$, however, went on to produce one more clutch, nest H. Both males obtained mating access and, in the absence of any other breeding activities in their system, both were free to help full-time in chick feeding.

Within this one mating system, therefore, there was huge variation in how males shared matings and parental care, with the degree of female synchrony playing a key role in how the alpha male was able to monopolise matings and in how males allocated care to broods. I shall now collate all the data on such sequences in polygynandry to summarise how alpha and beta males allocate their mating and parental effort between females.

Mating versus other options

When confronted with two females at different stages of the reproductive cycle, alpha males always preferred mating opportunities over all other options (Table 11.3). Beta males also preferred matings but they were more likely to forgo mating competition to help feed another female's brood[1] or to associate with another female at an earlier stage of the pre-laying period.[2] Given that alpha males are able to monopolise a greater share of the matings, competing for matings is less profitable for beta males which may explain their more frequent choice of the alternative options in Table 11.3.

Choosing between two mating females

Eggs are fertilised 24 hours before they are laid and there is likely to be second male sperm precedence (Chapter 12), so it would clearly pay a male to prefer the female who was closest to laying. Figure 11.4 shows that alpha males did indeed show a preference for the more 'valuable' female when confronted with a choice of two females who were both soliciting matings. However, when females were synchronous and both in the period 2 days before laying through to clutch completion, the alpha male showed no strong preference and tended to spend time with each (Table 11.3, Fig. 11.4). As a result, the beta male gained free access to whichever female the alpha male left unguarded and was thus more likely to gain a share of the mating access when two females were

Table 11.3. Reproductive allocation by alpha and beta males in polygynandry when faced with the choice of associating with a laying female (2 days before laying of the first egg to the end of egg laying) versus another female at various reproductive stages (from Davies and Hatchwell 1992).

Alternative option	Male	No. males who spent time:		
		with laying female	with alternative option	with both
Pre-breeding or	Alpha	10	–	–
nest-building female	Beta	6	1	1
Female with completed	Alpha	15	–	3
nest, more than 2 days	Beta	3	5	10
from laying of first egg				
Another laying female:	Alpha	2	–	7
from 2 days of laying of first	Beta	–	2	7
egg to end of egg laying				
Incubating female	Alpha	24	–	–
	Beta	19	–	–
Female with brood of young	Alpha	28	–	–
	Beta	11	9	–

In some cases, detailed time budgets were obtained only for the alpha male.

synchronous (26 out of 32 cases) than when they were asynchronous (33 out of 71 cases)[3] and the alpha male had the opportunity to monopolise each female in turn.

Choosing between two broods of young

When two polygynandrous females have synchronous young, Table 11.1 suggests that on average more young will be raised in total if each male helps full-time at one nest (payoff is $2 \times 2.41 = 4.82$) rather than both helping full-time at one nest and leaving the other female unaided (payoff is $3.00 + 0.81 = 3.81$), or both helping part-time at both nests (payoff $= 2 \times 2.00 = 4.00$). Figure 11.2(b) compares the payoffs for various combinations of brood sizes in the two nests and confirms that in the majority of cases the two males do best by each working full-time at one nest. Only when the choice is between broods of 5 v. 1, 4 v. 1 and 3 v. 1 do males do better by both helping the larger brood and leaving the female with the single chick unhelped. This effect arises simply

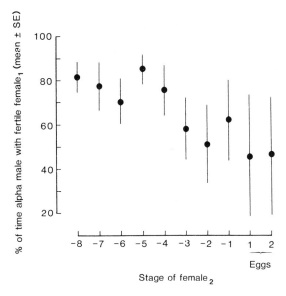

Fig. 11.4 Proportion of time an alpha polygynandrous male spends with one female (female$_1$), who is in the peak period for fertilising eggs (from 2 days before laying of the first egg through to the end of egg laying), in relation to the mating stage of another female, female$_2$, indicated by the number of days before the laying of the first egg. Data from 17 different males in 24 breeding attempts. Spearman rank correlation = −0.855, P < 0.01. From Davies and Hatchwell 1992.

because a male's help increases reproductive success more when added to a male plus female feeding a larger brood than when added to an unaided female with a single chick (Fig. 11.1).

Before we can predict how the two males should allocate their care, however, we also need to consider their mating access. As shown in Chapter 7, a male will help to feed a female's brood only if he has gained a share of the mating access with that female (Table 7.1). Because mating access is a prerequisite for chick feeding, I shall just consider those cases where a male has mated with two females and so has a choice of which brood to feed.

In 12 cases, both males mated with both females so both could have fed at either nest. In 6 cases, each male fed full-time at different nests. In one case, both males fed at both nests but the alpha male brought 77% of his feeds to one nest while the beta brought 67% of his to the other. In 3 cases, the beta fed at just one nest and the alpha brought 71%, 79% and 86% of his feeds to the other. Summarising, in a total of 10 cases each male had a strong or absolute preference for different nests and in only 2 cases did both males feed at one nest and leave the other female unaided.[4]

In 7 other cases, one male mated with just one of the females and fed only her brood, while the other male mated with both females and so had a choice of which brood to feed. In 6 cases, he fed full-time at the other nest, while in one case he fed at both nests but there were no data on his degree of preference.

Therefore, in at least 16 of the 19 cases, each male had a strong or absolute preference for different nests,[5] as predicted if they allocated their efforts so as to maximise the total number of young raised from the two broods. There were insufficient data to test the prediction that at certain splits of brood sizes, both males should feed full-time at one nest (Fig. 11.2b).

What determines which brood will be fed by the alpha male and which by the beta male? Of the 16 cases (above) where each male had a strong or absolute preference for different nests, there were no differences in brood sizes attended by alpha males (mean ± SE = 3.21 ± 0.31) versus beta males (3.00 ± 0.24),[6] nor in hatching order (in 6 cases the alpha fed the first brood to hatch, in 7 cases the beta did so and in 2 cases hatching occurred on same date; one case unknown).

However, mating access was a strong predictor of which brood a male chose. Figure 11.5 shows that a male put all, or most, of his effort into provisioning the brood of the female with whom he had gained a greater share of the mating access. In 25 of the 28 cases a male put all his effort into one brood, and in 24 of these cases he chose the brood where he had gained a greater share of mating access with the female during the mating period. In all 3 cases where a male

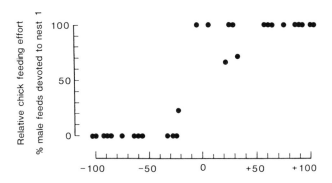

Fig. 11.5 How a male allocates his feeds between two synchronous broods of young in relation to his share of mating access with the two females. Positive values on the *x*-axis indicate that a male gained a greater share of the matings with female$_1$ than he did with female$_2$; negative values indicate he got a greater share with female$_2$ than female$_1$. Data from 14 cases of polygynandry (14 points for alpha, 14 for beta males) where each male gained mating access with at least one female, and where there were detailed quantitative data on mating share and chick feeding effort. From Davies and Hatchwell 1992.

divided his effort between two broods, he put a greater share of his effort into the brood of the female with whom he had enjoyed greatest mating access. This strong preference makes good adaptive sense, because DNA fingerprinting showed that share of matings reflected share of paternity (Chapter 12). Thus, each male tended to put all his effort into the brood where he had greatest paternity.

Although share of mating access influenced which brood a male chose, it did not influence how hard he worked at chick feeding (Fig. 11.6). When a

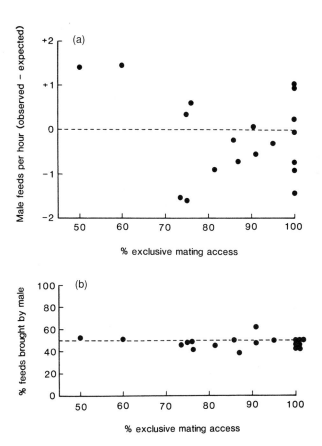

Fig. 11.6 Provisioning effort in relation to share of mating access for cases where a single polygynandrous male helped to feed the young. (a) Feeds per hour to day 5–12 nestlings in relation to the expected values for a monogamous male for the relevant brood size (from Fig. 10.2b). (b) The proportion of feeds (male plus female) brought by the male. The dashed lines indicate the expected values if there was no difference in work rate or work share from that of a monogamous male, who gains all the mating access and all paternity. From Davies and Hatchwell 1992.

polygynandrous male was the sole helper of a female, he provisioned at the same rate as a monogamous male, bringing on average 50% of the total feeds by the pair, irrespective of whether he had gained all of the mating access or only a share of the matings. This result echoes the finding from the previous chapter, that beta males work just as hard as monogamous males when alpha males are removed from polyandrous trios, even though they should be able to assess that they cannot have full paternity of the brood. The influence of paternity on parental effort will be discussed further in Chapter 12.

11.5 How best to allocate care between broods

In theory the relationship between amount of care and reproductive success is likely to be an S-shaped function of the form in Fig. 11.7 (e.g., Parker and Macnair 1978). At low levels of care, increased care is expected to have an accelerating effect on success. For example, when chicks are starving an increase in provisioning rate is very beneficial and may make the difference between life and death. Beyond a certain limit, however, there will be diminishing returns on increased parental care as provisioning begins to saturate the brood's needs. A well-fed chick will benefit very little, if at all, from yet more food stuffed into an already full gape. The beginnings of this decelerating region of the curve can be seen in the dunnocks, where a male's help increases success more when added to an unaided female than when added to a pair (Fig. 11.1). Thus, the difference between the help of one and two adults has a much greater effect on reproductive success than the difference between the help of two and three adults.

Figure 11.7 illustrates the consequences of this result for how males should best allocate their effort between two broods of equal size. Over the initial part of the curve increased investment has an accelerating effect on reproductive success. Thus if, without his help, each female can raise N_1 young, a male gains more from adding all his effort X to one female's brood than from adding half his effort X/2 to each of two broods, because $N_1 + N_3 > 2N_2$. This may represent the choice facing a polygynous male dunnock, where full-time help at one nest is usually most profitable. With further male help beyond an amount X, the increase in reproductive success is decelerating. Thus a second male gains more from adding all his care to an unaided female than from adding it to a brood already cared for by a female and male, because $2N_3 > N_1 + N_4$. This represents the choice facing two males in polygynandry, where each male usually cares for a different brood.

A game theory model by Peter Sozou and Alasdair Houston (in prep.) shows that under most circumstances it is stable for each male to help full-time at different nests, and for each to choose the brood where he has greatest paternity, just as observed.

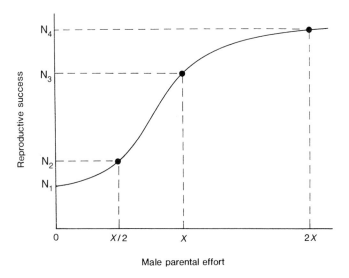

Fig. 11.7 Theoretical relationship between male parental effort and reproductive success; see text for discussion. From Davies and Hatchwell 1992.

Ecological conditions will often determine the value of extra parental care. At one extreme, under harsh conditions, the increase in reproductive success may not reach the decelerating part of the curve and so it may pay additional helpers to add their full-time care even to broods already attended by several adults (Emlen and Wrege 1988). At the other extreme, even unaided females may operate in the decelerating region so that it would pay helpers to distribute their aid among several nests or even to leave females to care alone (e.g., some fruit and seed-eating species, Lack 1968). Dunnocks provide an intermediate case where allocation is predicted to vary depending on the amount of help a female is already receiving. The main conclusion from this chapter is that given these various choices males distribute their care in a way which is consistent with maximising their reproductive success.

11.6 Summary

The variable mating system enables measurements of how different amounts of male help influences reproductive success (from no male help through one male part-time, one male full-time to two males full-time). Male help became increasingly valuable as brood size increased, but the addition of one male's help increased success more than the addition of a second male's help.

The data on the value of male care can be used to predict how males should best allocate their parental effort between two synchronous broods. As predicted: (a) a polygynous male preferred to invest full-time in one brood rather than part-time in both; and (b) in polygynandry, where two males shared matings with two females, each male provisioned at different nests rather than both helping one brood and leaving the other unaided.

In polygynandry, a male chose to feed the brood of the female with whom he had gained a greater share of the matings (and hence paternity). However, mating share did not influence the provisioning rate of a lone male helping a female; he worked equally hard whether he gained all or just a share of the matings.

Males preferred mating to chick feeding and, with two fertile females, spent most time with the female closest to laying.

The main conclusion is that with choices involving trade-offs between two mating efforts, two parental efforts, or mating versus parental effort, males chose the option which maximised their own reproductive success.

STATISTICAL ANALYSIS

1. Table 11.3; $\chi^2_1 = 12.69$, $P < 0.001$.
2. Table 11.3; $\chi^2_1 = 13.44$, $P < 0.001$.
3. $\chi^2_1 = 9.52$, $P < 0.01$.
4. Binomial test, 2-tailed, $P = 0.038$.
5. Binomial test, 2-tailed, $P = 0.004$.
6. $t_{15} = 0.456$, NS.

12

Paternity and parental effort: how good are male chick feeding rules?

12.1 Do dunnocks have to be clever to behave adaptively?

One of the most important findings of this study is that male and female behaviour makes good adaptive sense in terms of maximising individual reproductive success (Chapter 9). The first reaction of a colleague was: 'Well, those dunnocks must be very clever to work out all the reproductive payoffs in Table 9.4!'

The standard reply is that the animals themselves need not be aware of the consequences of their actions. The dunnock's cryptic plumage has presumably evolved without individuals being aware of exactly how the crypsis works. In the same way, behavioural strategies can evolve without individuals working out all the costs and benefits. It is natural selection which judges the success of alternative strategies, favouring those which best promote an individual's reproductive success, the means by which genes programming those strategies proliferate in the population. Individuals may play their strategies unconsciously, as programmed by their genetic and physiological make-up; this will influence, for example, plumage colour, the tendency to be aggressive to rivals, work rate in chick feeding, and so on.

It has to be admitted, however, that rather little is known about how these alternative behaviour patterns might be produced at the mechanistic level. Recent studies have suggested that mating systems may be regulated proximately by temporal patterns of hormone secretion. For example, in polygynous species, males have high levels of circulating testosterone for longer periods than in monogamous species. Implants in normally monogamous species lead to reduced parental care by males and increased territory size, so inducing polygyny (Wingfield 1984; Hegner and Wingfield 1987). Thus, testosterone levels in males may represent a compromise between the advantages of allocating effort to caring for young versus gaining extra mates.

Although, in principle, there may be relatively simple physiological mechanisms underlying a male's choice of mating and parental strategies, several results in previous chapters suggest that dunnocks are doing more than just blindly obeying the predicates of their physiological state. For example, a male may copulate with two females but help only one of them with offspring care. Furthermore, males may switch suddenly from feeding the offspring of one female to competing for matings with another female, or switch strategy in response to a change in behaviour by a rival male (Chapter 11). These choices give the impression that males are indeed somehow assessing the costs and benefits of different reproductive opportunities, just as individuals can assess the profitability of alternative foraging strategies (Krebs *et al.* 1977, 1978). If it seems too much to expect the birds to work out the detailed reproductive pay-offs from each course of action (Table 9.4), then presumably they must use some simple rules to guide their behaviour, in the same way that a foraging animal can use simple mechanisms to determine how good a feeding patch is (Stephens and Krebs 1986; Schmid-Hempel 1986).

The aim of this chapter is to discover something about the rules used by males to allocate their parental effort. Do they have some equivalent of DNA fingerprinting to assess their paternity? If not, what rules do they use to guide their parental effort? How good are those rules? The chapter is based on some field observations and experiments I did together with Ben Hatchwell in the three summers of 1988 to 1990, combined with an analysis of paternity using DNA fingerprinting done by Terry Burke and Tim Robson (Davies *et al.* 1992).

12.2 Influence of natural variation in mating access on paternity and chick feeding

As discussed in Chapters 6 and 7, there was intense competition for matings between alpha and beta males in polyandry and polygynandry and the outcome was very variable. Sometimes, alpha males guarded females closely throughout and beta males were never seen to gain access. In other cases, access was shared between alpha and beta male but the share varied widely; although alpha males usually gained more, occasionally beta males did better, especially on territories where vegetation was dense and where alpha males found it difficult to maintain close contact with the female. In some cases, both males lost the female for long periods and she hid away feeding alone quietly, while in others, one or both of the males escorted her almost all of the time.

This variation provides a valuable source for examining how a male's share of matings influences his share of parental effort. We have already seen (Chapter 11), that when a polygynandrous male shares matings he nevertheless provisions at the same rate as a monogamous male, who has full paternity, if he is the sole male helping a female to feed the brood (Fig. 11.6). How does share of matings influence share of parental effort where two males are free to help at the same nest? This includes all cases of polyandry (where the two males have just one female) and those cases of polygynandry where the males are not affected by the activities of other females in their system (i.e., excluding all cases where other females were laying or had young at the same time).

A male's share of matings has been scored as the proportion of 'exclusive access time' he gained with the female, time when there was only one male within 10m of the female, when copulations could therefore proceed un-interrupted (Chapter 6). Male mating success was scored as share of mating time rather than share of copulations simply because the birds often hid away in dense vegetation so copulations themselves would not always be seen. However, share of access time certainly gave a good measure of share of copulations because there was no difference in alpha versus beta male copulation rate during periods of exclusive access (Section 7.4), nor was there any variation in their share of access at different stages of the mating period (Section 6.5).

Figure 12.1 includes only those cases where both alpha and beta male gained some exclusive mating access and shows that a male's share of parental effort increased with his share of exclusive access to the female during the mating period. Figure 12.2 shows, furthermore, that an increased share of the mating access led to an increased share of the paternity of the brood. These two results suggest that a male does not simply have a 'feed versus do not feed' rule based on mating access (Chapter 7), but varies his share of parental effort in relation to his probability of paternity. These data raise two interesting problems.

(a) First, is a male really monitoring his paternity share by assessing his mating access relative to that of a competing male, as suggested in Fig. 12.1, or is he only assessing his own total access? These two alternatives might suggest different physiological mechanisms. If a male responded just to his own amount of access then the mechanism could be a simple one; for example, increased copulations may cause a change in the male's physiological state (e.g., circulating hormone level) which in turn could set the level of investment in parental care.

In fact the evidence argues against a mechanism based on a male assessing only his own mating success. The proportion of the *total* time that a male gains exclusive mating access is indeed correlated with share of parental effort, but the correlation is weaker than that of *share* of exclusive access. Furthermore, when partial correlation is used to control for the effects of the other variable, proportion of total time becomes a weak predictor of share of effort while share of access remains strongly correlated (Davies *et al.* 1992). This suggests that a male monitors his paternity share by comparing his exclusive mating access with that of his rival, rather than by just assessing his own amount of access. The mechanism involved must, therefore, be more complex than the simple one envisaged above.

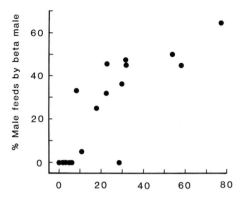

Fig. 12.1 A beta male's share of the male feeds during chick provisioning increased with his share of the matings, as measured by his share of the exclusive access to the female during the mating period. Correlation coefficient (log-transformed data), $r = 0.775$, 15df, $P < 0.001$. Data are from natural variation in alpha:beta share of mating access in polyandry and those cases of polygynandry where the provisioning of the brood was not constrained by the activities of other females in the mating system. In all 17 cases here the beta male gained some exclusive access to the female. From Davies *et al.* 1992.

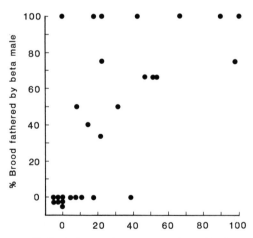

Fig. 12.2 A beta male's share of the paternity of a brood, assessed by DNA finger-printing, increased with his share of the exclusive mating access with the female; Spearman rank correlation (corrected for ties), $r_s = 0.685$, $n = 28$, $P < 0.01$. For the 20 cases where the beta male obtained some mating access, $r_s = 0.665$, $P < 0.01$. From Davies *et al.* 1992.

(b) The relationship in Fig. 12.1 is based on natural variation and so may not be a causal one. For example, those beta males who are better competitors may both gain more mating access and be able to invest more in offspring care. In the next section, therefore, an experimental approach is used to test whether increased mating share leads to increased parental effort. The experiments also provide another advantage. Although, on average, there was no significant variation in the share of exclusive access gained by alpha and beta males at different stages of the mating period, matings at different stages may vary in their chances of fertilising eggs and we were interested to test whether males placed different value on copulations achieved at different times in relation to egg laying.

12.3 Influence of experimental variation in mating access on chick feeding

The mating access of monogamous males ($n = 15$) and alpha males in polyandry and polygynandry ($n = 28$) was varied experimentally by removals of 3 days' duration at various stages during the mating period (Fig. 12.3). The males, caught in mist nets, were kept in aviaries away from the Garden and fed on Claus soft-bill food. All maintained good weight and all settled back quickly on their original territories when released. The aim of the experiment was to use male chick feeding responses to tell us how the males themselves assessed the link between matings and paternity, and to use the DNA fingerprinting to tell us how mating access actually influenced paternity. By comparing the two, we could then test how well a male's chick feeding behaviour promoted his own reproductive success.

When monogamous males were removed, neighbouring males came onto the territory and mated with the female. When alpha males were removed, resident beta males took over control of the female and were just as successful in keeping neighbours at bay as when the alpha male was in charge (Chapter 6). It might be thought that these experiments, giving males windows of access at different stages of the mating period, placed the birds in an unnatural situation. However, we have seen such variation occurring naturally, earlier in the book. For example, in some cases of polyandry, beta males gained access only at particular stages, and in polygynandry beta males could be allowed free access at various times depending on when the alpha male was enticed away by the activities of other females in the system (Chapter 11).

All males removed during incubation (controls), after the mating period had ended, gained normal mating access both before and during egg laying and all fed the young (Table 12.1). Thus, the experience of the temporary removal itself did not disturb the male's normal chick feeding behaviour. Males who were removed at stages C in Fig. 12.3, who thus had reduced mating access but who

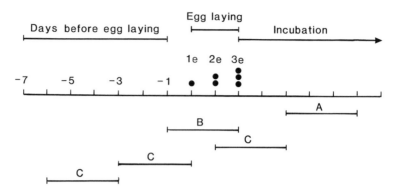

Fig. 12.3 Diagram to illustrate the stages of experimental removals of monogamous males and alpha males. In this example, the female lays a clutch of 3 eggs (1 egg per day) and begins incubation on the day the last egg is laid. The temporary removals, for 3 day periods, were at various stages indicated by A–C. Males removed at stage A (incubation) gained normal mating access. Males removed at stage B gained mating access prior to egg laying but not on days that eggs were laid. Males removed at stage C had reduced mating access but gained at least some access during egg laying. From Davies *et al.* 1992.

were present during at least some days of egg laying, likewise all fed the young. By contrast, of the seven males who were removed throughout the entire egg laying period only one fed the young, even though all seven had enjoyed several days mating access prior to removal and even though all seven settled back on their territories (Table 12.1).

Now, considering the responses of the beta males who gained access as a result of alpha male removal, 11 gained access both before and during laying and all helped to feed the young, while 13 gained access during laying only and all 13 likewise fed the young.

These results suggest that some mating access during egg laying itself is necessary to cause chick feeding. Even a single day's access during egg laying was sufficient; eight of the removed males gained access only on the day the first egg was laid and three gained access only on the day before the last egg was laid, yet all eleven fed the young. Likewise, of the beta males who gained increased mating access due to the alpha male's removal, six gained access only on the day before the last egg was laid yet all six helped to feed the young.

The experimental removals provided no evidence for infanticide by males (see Chapter 8) that were removed throughout the egg laying period (Davies *et al.* 1992). Males that were removed throughout egg laying had nevertheless enjoyed mating access for up to a week prior to removal, so removals of longer than 3 days would be needed to reduce mating access to zero. It is possible that there are two thresholds of access influencing male behaviour, with some access being

Table 12.1. Influence of stage of experimental removal on whether males fed the young (from Davies *et al.* 1992).

Stage removed (see Fig. 12.3)		Mating access to female	No. removed males who fed young		
			Monogamous males	Alpha males	Total
Incubation	(A)	Before and during laying	7/7	4/4	11/11*
Mating period	(B)	Before laying only	0/2	1/5	1/7†
Mating period	(C)	Before and during laying	6/6	19/19	25/25

* Comparing totals in A and B, $P < 0.01$ (Fisher exact tests). † Comparing totals in B and C, $P < 0.001$.

sufficient to prevent infanticide, but access during laying itself being necessary to cause chick feeding.

12.4 Do males use the onset of laying to value their copulations? An experiment with model eggs

Males were clearly interested in the nest contents and inspected the nest frequently during the mating period. When we released males back on their territories after the removals, one of the first things they did was to visit the nest. Ben Hatchwell and I used an experiment to test whether males used the appearance of an egg in the nest as a cue to value their copulations.

Once females had completed nest-building, were being guarded and were copulating, we placed a model dunnock egg in the empty nest. Model eggs, made of gel coat resin and painted to match the bright blue of real dunnock eggs, were 'laid' in the early morning, to mimic the time of laying of real eggs. After one day with the model in the nest, we attempted to catch the guarding male before real laying began. We were not always successful on the first day, in which case we tried again on the following days, though if real laying began then the experiment had to be abandoned. As soon as the male was caught, the model was removed from the nest and the male was then kept in an aviary, away from the Garden, until the female had completed her clutch and had been incubating for one day. The male was then put back on his territory and we waited to see whether he would feed the chicks when they hatched.

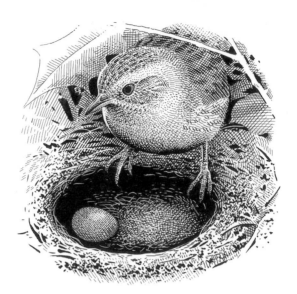

Males often visit the nest during the mating period. An experiment with model eggs demonstrates that the sight of an egg is one of the cues used by males to value their copulations. Males were more likely to feed the chicks if they had gained copulations during the time that eggs appeared in the nest.

Of those males removed within 24 hours of the laying of the first egg, and kept in captivity throughout real-egg laying, only one out of six controls (no model egg) fed the young, even though they had all enjoyed mating access for several days prior to removal. By contrast, all seven males who experienced a model egg before removal subsequently fed the young, a significant difference[1] (Hatchwell and Davies 1992*a*). Thus the sight of an egg in the nest is used as a cue by males to value their copulations. It is not the only cue, however, because of the males who experienced a model egg but who were removed two or more days before egg laying, none out of four fed the nestlings.[2] Thus the presence of an egg in the nest must be combined with female appearance or behaviour indicative of imminent laying if it is to cause chick feeding by males. As shown in Chapter 6, males seemed to be able to detect when females were about to lay because competition for matings increased the day before egg laying began.

12.5 Are male responses adaptive? Influence of removals on paternity

The results in the previous two sections show that a male demands mating access during laying if he is to 'value' his copulations, in the sense of investing in chick

Table 12.2. How male removals at different stages of the mating period affected a resident male's paternity, assessed by DNA fingerprinting. The table considers only cases where observations indicated that one resident male monopolised all the matings throughout the mating period (controls) or prior to removal (experimentals) (data from Davies *et al.* 1992).

Treatment	Resident male's paternity	
	% broods (*n*) with full paternity	% young sired (*n*)
(a) *Controls*		
No removal	96.6 (29)	98.9 (88)
· Removal after clutch completed	80.0 (5)	91.7 (12)
Total	94.1 (34)	98.0 (100)
(b) *Experimental removals*		
Before first egg laid	10.0 (10)	23.1 (26)
Day first egg laid	20.0 (5)	50.0 (16)
After day of first egg, but before clutch completion	100 (12)	100 (32)
Total	51.9 (27)	62.2 (74)

feeding. We can test whether this behaviour is adaptive by examining how the removal experiments affected paternity.

Table 12.2 includes only those cases where observations showed that one male had monopolised all the mating access prior to removal, in other words, monogamous males and those alpha males in polyandry and polygynandry who had been successful in keeping beta males at bay. The control data (removals after clutch completion, when all the eggs have been fertilised, and no removals) show that our observational assessment of male monopolisation was quite accurate. Of the cases where we scored one male as gaining all the matings, only 2 out of 100 chicks (one chick in each of two out of 34 broods) were sired by another male. One was sired by a neighbour but we were unable to test the other against potential sires because of insufficient sample for DNA analysis.

Compared with these controls, experimental removals (before and during egg laying) clearly increased the chance that another male would father the young.[3] The resident male lost paternity in 13 of the 27 broods. In all cases where an alpha male was removed (from polyandry or polygynandry), it was the beta male who took over the female and gained paternity at the alpha male's expense. In removals of monogamous males, it was a neighbour who took over the female and gained paternity (Davies *et al.* 1992).

Table 12.2 also shows that, within the experimental removals, the stage of removal influenced the degree of paternity loss,[4] with earlier removals during the mating period increasing the chance that the replacement male gained paternity. Indeed, although all removed males had enjoyed mating access for several days before removal, provided they were removed before the first egg was laid the replacement male sired most of the brood. Such 'second male advantage' has been shown previously by Tim Birkhead and his colleagues in laboratory experiments with zebra finches *Taeniopygia guttata* (T.R. Birkhead *et al.* 1988). They found that even a single copulation by a replacement male had a very high chance of leading to fertilisation of subsequent eggs. The mechanism creating such 'last male sperm precedence' is not known, but one possibility is that the sperm from the males becomes stratified in the female's sperm storage tubules (Fig. 6.16), with the second male's sperm overlaying the sperm from previous matings. The second male's sperm may then be more likely to fertilise eggs through a process of 'last in—first out'.

It is not possible to tell from our experimental results with the dunnocks whether last male sperm precedence accounts for the replacement male success, because number of copulations by removed and replacement males may also have varied. Thus, replacement males may have been more likely to fertilise eggs because of last male advantage or because they gained more matings at a critical time, or by a combination of these two. Whatever the mechanism, however, it was particularly pleasing to find this effect in the dunnocks because it makes sense of two observations discussed in Chapter 6, namely the continued competition for matings throughout egg laying and the tendency of a male to copulate as soon as he encountered a female, after a period when another male had been with her. Second male advantage may also have been expected from the dunnock's extraordinary pre-mating display; although the ejected sperm may not originate from the female's sperm storage tubules, but rather from the cloaca and vagina, the display must presumably enhance the chances of the replacement male gaining paternity (Chapter 6).

These paternity results show why it pays males to refuse to feed the brood if they do not gain matings up to the onset of laying. Males removed before laying have no definite marker by which to assess their chances of paternity. Given that females solicit matings several days before the laying of the first egg and given the fact that replacement males are quick to take over unguarded females and have high success at fertilising eggs, it makes good sense that a male removed before egg laying, and absent throughout laying, refuses to feed the chicks despite his previous mating access—he has a high chance of not fathering any of the brood. By contrast, males present up to the start of egg laying are guaranteed some paternity provided they have monopolised matings up to then, so their provisioning of the brood makes good adaptive sense.

Table 12.2 suggests that males who gain access only later on in laying, after the day of the first egg, are unlikely to achieve much success. Males who gain access only on the last day before clutch completion have little or no chance of fertilising eggs because the last egg will have been fertilised at dawn that day (eggs are fertilised 24 hours before they are laid). Copulations throughout the last day are therefore worthless, yet males continue to copulate and compete for matings right up to the onset of incubation (Chapter 6). A likely explanation is that males are not certain that the following morning's egg is the last one (Chapter 6). However, once the female begins incubation it would clearly pay males to backdate and devalue the previous day's copulations. They clearly do not do this—all 9 males who gained mating access only on the last day before clutch completion, when there were no more eggs to fertilise, nevertheless fed the young (see above).

Thus, the male's chick feeding rule of 'feed the young if I gained mating access during egg laying', makes good sense in relation to the greater loss of paternity to replacement males early on in the mating cycle, but it is not completely foolproof. There are two drawbacks. First, the rule leads males to undervalue matings achieved before laying, which could fertilise the first eggs in the clutch. Second, it leads them to overvalue later matings, achieved at a time when most, or even all, of the clutch is already fertilised.

In conclusion, these experiments reveal that males use simple rules to guide their parental effort, based on mating access to predict paternity and the appearance of eggs in the nest to mark the stage at which copulations are likely to be 'valuable' in the sense of siring offspring. Although the DNA fingerprinting shows that the rules are not perfect, they work reasonably well in getting males to expend more parental effort in cases where they are likely to have greater paternity.

12.6 Does share of matings determine parental effort?

I now turn from the male's decision of whether or not to feed the young to the question of how hard he should work once he has decided to provision. Because the experiments indicated that males valued mating access only during laying, chick feeding effort has been examined in relation to access on egg-laying days.

Alpha versus beta male share of provisioning in trios

Figure 12.4 plots the provisioning rates of alpha and beta males in relation to their experimentally varied share of the mating access, caused by the alpha male removals at different stages of the mating cycle. Alpha males worked harder the greater their share of the matings. The beta male's provisioning rate varied with his share of the mating access in exactly the same way, being greater when alpha males were removed for more of the egg laying period, thus allowing beta males an increased share of the access at this time.

Figure 12.5 expresses these experimental results as the alpha male's share of the male feeds to the brood in relation to his share of the mating access during

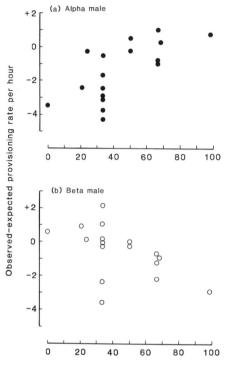

Fig. 12.4 Provisioning rates of nestlings by alpha and beta males in polyandry and polygynandry, where both males helped to feed the brood, in relation to their experimentally varied share of the mating access during egg laying, caused by 3 day removals of alpha males ($n = 17$) at various stages of the mating cycle. Provisioning rates are observed rates per hour in relation to those expected from average alpha and beta provisioning rates to a brood of a given size (Chapter 10). (a) Alpha males: Spearman rank correlation, $r_s = 0.667$, $P < 0.01$. (b) Beta males: $r_s = -0.553$, $P < 0.05$. Expressing a male's provisioning rate in relation to his own share of the mating access (arcsine transformed access data); for alpha males, $y = -4.05 + 0.0642x$, and for beta males $y = -3.15 + 0.044x$, ANCOVA shows no difference in slope ($F_{1,30} = 0.57$) or elevation ($F_{1,31} = 0.05$). Thus, both males increased work rate in relation to mating share in the same way. From Davies *et al.* 1992.

laying. The result echoes that found earlier from natural variation (compare it with Fig. 12.1), supporting the view that a male's share of provisioning effort in a trio is determined by his share of matings (and hence paternity).

When alpha males were released back on their territories following the removal, there was a brief period of chasing and fights between the two males. Sometimes the original dominance order was restored, with the removed bird

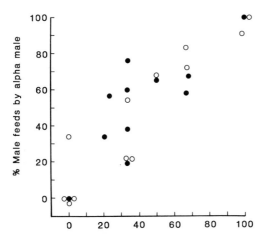

Fig. 12.5 Relationship between an alpha male's experimentally varied share of mating access (caused by removals at different stages of the mating period) and his share of male feeds brought to the nestlings. Points at (0,0) on the graph indicate that the beta male gained all the mating access and did all the chick feeding. Points at (100,100) indicate that the alpha male gained all the mating access and did all the chick feeding. Other points indicate that both matings and chick feeding were shared between the two males. For all 23 points (including cases where one male provisioned and both males provisioned), Spearman rank correlation, $r_s = 0.869$, $P < 0.001$. For the 17 cases where both males fed, $r_s = 0.702$, $P < 0.01$. In some cases the removed male retained alpha status when released back on his territory (●), in other cases he lost status and became beta (○). ANCOVA showed that there was no difference between these two cases (arcsine transformed data: slope $F_{1,19} = 0.01$; elevation $F_{1,20} = 0.34$). From Davies *et al.* 1992.

retaining alpha status but sometimes the dominance order reversed and the original alpha became beta (see Chapter 5). Figure 12.5 shows that there was no effect of dominance status on chick feeding share; the relationship between provisioning share and mating share was the same whether the removed male maintained rank or lost rank. This provides nice evidence that it is mating access which influences chick feeding, not dominance rank *per se*. When beta males were experimentally given a greater share of the mating access, they did a greater proportion of the male feeds irrespective of their dominance rank later on at the time of chick feeding.

Monogamous male effort in pairs

How do alpha and beta males come to their seemingly fair split of work in relation to their share of matings (Fig. 12.5)? We think some kind of bargaining

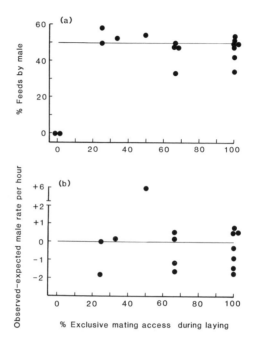

Fig. 12.6 Chick provisioning by monogamous males in relation to their experimentally varied mating access during egg laying, caused by 3 day removals at different stages of the mating period ($n = 17$). In all cases the removed male was the only male available to assist the female in chick feeding. (a) Percentage total feeds by male. The line is the expected 50% for normal cases of monogamy. (b) Observed male provisioning rate per hour in relation to that expected from a monogamous male for a brood of a given size (Chapter 10). The line is the expected relationship if the male fed at the same rate as an unmanipulated monogamous male. Provided the male gains some mating access during laying, there is clearly no influence of amount of access on chick feeding effort. From Davies *et al.* 1992.

must occur between the males because our removals of monogamous males produced a different result.

Although the monogamous removals led to increased mating access by neighbours, in almost all cases the neighbours went back to help their own females during chick feeding. Thus, the removed monogamous male was the only male available to help the female. Figure 12.6 shows that provided he gained some access during egg laying his work rate in chick provisioning was unaffected by share of mating access. Access on just one egg laying day caused the same provisioning rate as access throughout egg laying despite the paternity loss revealed by the DNA fingerprinting. In all cases the male did around 50% of the feeds to the nest, as is normal for paired males (Chapter 10).

12.7 Why does paternity loss influence male work rate in trios but not in pairs?

Why does share of mating access influence alpha and beta male effort in trios but not monogamous males in pairs? One simple explanation could be a difference in knowledge. Alpha males could easily assess that they have lost paternity because when they return to their territory they find the beta male in residence and later see him feeding the chicks. By contrast, because neighbour replacements return to their own territories, when a monogamous male is released he may simply find the female incubating and so has no way of assessing whether he has lost paternity during his absence.

However, there are three reasons for dismissing this as the explanation. First, in some cases the monogamous male was returned before clutch completion, and so in time to find a neighbour with his female. Second, when an alpha male is removed during chick feeding from a territory where both males are helping the female to feed the brood, the beta male immediately increases his provisioning rate to that of a monogamous male with full paternity (Chapter 10). He could surely assess both from the fact that matings were shared and chick feeding was shared that he did not gain full paternity. Third, in polygynandry, alpha and beta males often share paternity of two synchronously laying females and then each male helps to feed one of the broods. The male's work rate is not related to mating share but is the same as that of a monogamous male with full paternity (Chapter 11).

I suggest the following explanation. In theory, a monogamous male should respond to paternity loss in exactly the same way as he does to a reduction in brood size, namely by reducing his provisioning rate (Chapter 10). However, these two cases are not equivalent in practice because a male apparently cannot recognise his own young in a multiply-sired brood (Chapter 7). If a lone male decreased his effort in response to paternity loss, the whole brood would suffer, including those chicks sired by him. Certainly, females increase their effort a little in response to decreased male effort, but it is not sufficient to compensate fully for the male's reduction, so nestling weight and survival is reduced as a result (Chapter 10).

By contrast, when there are two males on a territory a male who loses paternity does so to another male helping at the same nest. Thus, any reduction in parental effort by one male in relation to his lost paternity is counteracted by an increase in effort by the other male, who has gained increased paternity. The similar relationships between paternity and parental effort for alpha and beta males (see Fig. 12.4) suggests that any reduction in effort by one of them is completely compensated for by an increase in effort by the other. Thus, whereas the chicks sired by a monogamous male would suffer if he reduced his effort in relation to paternity loss, the chicks sired by one of two polyandrous males do not. Further experiments are needed to reveal exactly how the alpha and beta males adjust their parental effort in relation to their share of paternity. They may each vary their parental effort not only in response to paternity, as suggested in Fig. 12.4, but also in response to each other's parental effort (see Chapter 10). It would be interesting to manipulate both these factors independently to see how the two males reach an 'agreement' over how much work to do.

To complete this story we also need to quantify how male care influences parental survival as well as nestling survival. In principle, there are circumstances when it may pay even a single male to reduce his effort in response to paternity loss because of the benefits of saved investment for future, more valuable broods. Previous studies have shown variable male responses to paternity loss through extra-pair copulations; in some cases males may reduce parental effort (Møller 1988*a*) while in others they apparently do not (Morton 1987; Westneat 1988). Male responses may vary depending on the benefit of provisioning to the current brood and the cost this has to the male's future reproduction (Houston and Davies 1985; Whittingham *et al.* in press).

12.8 How females might allocate matings to maximise male help in trios

The results in Figs. 12.4 and 12.5 show that there was no intrinsic difference in the abilities of alpha and beta males to provision young. The relationship between their chick feeding effort and their mating share was exactly the same for both (see the legend to Fig. 12.4). We have seen from earlier chapters that it clearly pays a female to gain copulations from both alpha and beta males so as to enjoy extra help with provisioning. Is there a particular share of provisioning which maximises total male help? And can females achieve this by allocating matings appropriately between the males?

Alpha and beta male provisioning rates followed the same curvilinear increase with share of male feeds, with the total male provisioning rate reaching a peak at around a 50:50 share of the feedings (Fig. 12.7). Given the relationship found between provisioning effort and mating share, the female would attain this maximum male help if she distributed the matings equally between the two males during the mating period. However, females were not free to allocate matings as they wished, because it payed alpha males to monopolise all the matings if they could (Chapter 6). The degree of female control over who gained matings was therefore largely dependent on her ability to escape the attentions of the guarding alpha male.

One measure of a female's ability to do this is the proportion of total observation time during the mating period that she was alone (no other individuals within 10m). Figure 12.8 indicates that when females were better able to escape the alpha male's close guarding, as measured by their time alone,

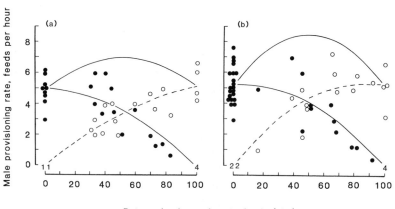

Fig. 12.7 Provisioning rates of alpha males (●) and beta males (○) relative to the beta male share of male feeds, for broods of (a) 2 chicks and (b) 3 chicks. Includes all cases of polyandry and those cases of polygynandry where the two males were not constrained by feeding activities at another nest or matings with another female in the system. Quadratic regressions provided significantly better fits than linear regression. Numbers at *x,y* 0,0 and 100,0 indicate number of data points. The top curve gives the total male rate calculated from the alpha and beta male curves. Peak total male rate occurs at around a 50:50 share of the provisioning. From Hatchwell and Davies 1990.

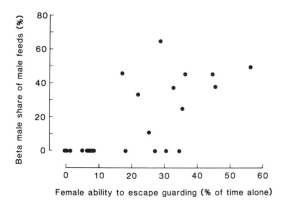

Fig. 12.8 Relationship between female ability to escape guarding by males (percentage of total observation time females spent alone during mating period) and the proportion of the male feeds provided by the beta male ($r_s = 0.634$, $P < 0.01$, $n = 18$). From Hatchwell and Davies 1990.

beta males were able to increase their access and contributed a greater share of the male feeds. There is considerable scatter in the data points, but there is some hint that at high levels of female 'control' the predicted 50:50 share of provisioning might be achieved.

12.9 Summary

In polyandry, natural variation in alpha versus beta male share of matings was correlated with their paternity share (assessed by DNA fingerprinting) and with their share of work in chick feeding. Thus, the two males varied their parental effort in relation to their probability of paternity. A male's *share* of the matings was a better predictor of his parental effort than was his *total* amount of matings, suggesting that a male might monitor his paternity share by comparing his mating access with that of his rival.

Males were removed temporarily at various stages of the mating cycle to create experimental variation in mating success. DNA fingerprinting revealed that replacement males sired most of the eggs fertilised during the removal period.

Removed males fed chicks only if they had gained matings during egg laying. This behaviour was adaptive because of the greater loss of paternity to replacement males earlier on in the mating cycle, but it led males: (a) to undervalue matings achieved before laying, which could fertilise the first eggs in the clutch; and (b) to overvalue later matings, achieved at a time when most or even all of the clutch was fertilised.

An experiment showed that males used the appearance of an egg in the nest as a cue to value their copulations. When a model egg was placed in a nest and males then removed within 24 hours of laying, they subsequently fed the young even though absent throughout real-egg laying. An egg was not the only cue, however, because a model egg did not stimulate chick feeding by males removed two or more days before real-egg laying. Thus female appearance or behaviour indicative of imminent laying was also important.

The removals confirmed that the alpha:beta share of parental effort in polyandry was determined by their share of matings, not by their dominance rank. By contrast, experimental manipulation of a monogamous male's mating access did not influence his parental effort. This difference is discussed.

Females attained maximum male help when two males shared provisioning equally. Females were more likely to achieve this when they were able to escape alpha male guarding and exercise some control over how the two males shared the matings.

STATISTICAL ANALYSIS

1. Comparing model egg results with controls; $P < 0.01$, Fisher exact test.

2. Significantly different from cases where males removed within 24 hours of laying; $P < 0.01$, Fisher exact test (Hatchwell and Davies 1992a).

3. Table 12.2: control versus experimental totals for proportion of broods with full paternity, $\chi^2_1 = 12.308$, $P < 0.001$; for proportion of young fathered, $\chi^2_1 = 35.810$, $P < 0.001$.

4. Table 12.2: considering the three stages of experimental removal; for proportion of young sired, $\chi^2_2 = 37.451$, $P < 0.001$; for proportion of broods with full paternity, $\chi^2_2 = 20.342$, $P < 0.001$.

13

Parasitism by cuckoos

13.1 Introduction

The main message of Chapter 12 was that dunnocks use simple rules to guide their parental effort. Females always feed the young in their nest and, in the absence of intraspecific brood parasitism (Chapter 7), they can have complete confidence that all the young dunnocks in their nest are their own. For males, the sharing of copulations reduces paternity but their use of mating access as a means of regulating their chick provisioning results in them directing most effort into broods where they have greater paternity.

In this chapter I shall discuss a case where the dunnock's simple rules lead to a maladaptive outcome, namely the rearing of a cuckoo chick. Given the finely tuned behavioural adaptations of dunnocks revealed in the previous chapters, their easy exploitation by cuckoos presents a real puzzle. Should we not expect hosts to evolve counter-adaptations to defeat cuckoos? If so, why is the dunnock apparently totally defenceless against this brood parasite?

13.2 The curious habits of the cuckoo

The cuckoo *Cuculus canorus* is a brood parasite, laying all of its eggs in the nests of various species of passerine birds, with the dunnock being one of its favourite hosts in Britain. The cuckoo relies entirely on the hosts to incubate its eggs and rear its young. The female cuckoo lays one egg per host nest, usually parasitising the nest during the host laying period. The cuckoo chick commonly hatches first, whereupon, just a few hours old and still naked and blind, it balances the host eggs one by one on its back and ejects them from the nest. Any newly hatched host young suffer the same gruesome fate, and so the cuckoo chick becomes the sole occupant of the nest.

The hosts then slave away, feeding the young cuckoo in the nest for about 20 days, and then for a further two weeks after fledging, until it becomes independent. This presents an extraordinary spectacle. When the nestling

The female cuckoo first removes one of the dunnock's eggs and, holding it in her bill, she then lays her own egg directly into the nest.

cuckoo is full grown it overflows the tiny nest, which may either disintegrate entirely or at best be flattened into a little platform. The hosts seem to risk being devoured themselves as they bow deep into the enormous gape with food. Once the cuckoo leaves the nest, the situation appears even more ridiculous as the hosts may have to perch on the cuckoo's back to reach the mouth of a fledgling eight times their own weight! Nevertheless, throughout chick-rearing the hosts behave as if nothing was amiss and as if they regarded the monster as one of their own young.

The parasitic habits of the cuckoo were known to Aristotle, writing some 2,300 years ago: 'it lays its eggs in the nest of smaller birds' (Peck 1970, p. 251); 'they . . . do not sit, nor hatch, nor bring up their young' (Hett 1936, p. 241). Aristotle also knew that the young cuckoo ejected the host eggs and young: 'when the young bird is born it casts out of the nest those with whom it has so far lived' (Hett op. cit.). However, it was Edward Jenner (1788) who first described the amazing performance of the young cuckoo in detail, having observed one in a dunnock nest found in June 1787. His marvellous description is worth quoting—in the subsequent 200 years it has not been bettered. Jenner uses the old name for the dunnock, 'hedge sparrow'.

'The little animal, with the assistance of its rump and wings, contrived to get the bird (the young hedge sparrow) upon its back and, making a lodgement for the burden by elevating its elbows, clambered backward with it up the side of the nest till it reached the top, where, resting for a moment, it threw off its load with a jerk and quite disengaged it from the nest. It remained in this situation a short time feeling about with the extremities of its wings, as if to be convinced whether the business was properly executed, and then dropped into the nest again. With the extremities of its wings I have often seen it examine, as it were, an egg and nestling before it began operations; and the nice sensibility which these parts appeared to possess, seemed sufficiently to compensate the want of sight which as yet it was destitute of. I afterwards put in an egg, and this, by a similar process, was conveyed to the edge of the nest and thrown out. The singularity of its shape is well adapted to these purposes: for, different from other newly hatched birds, its back from the scapulae downwards is very broad, with a considerable depression in the middle. This depression seems formed by nature for giving a more secure lodgement to the egg of the hedge-sparrow or its young one, when the young cuckoo is moving either of them from the nest. When it is about twelve days old this cavity is quite filled up, and then the back assumes the shape of nestling birds in general'.

The ejection behaviour is normally strongest from about 8 to 36 hours after hatching, and thereafter declines. If any host young remain after this time, they are usually smothered by the faster growing cuckoo and die anyway. Even while the foster parent broods it, the young cuckoo may continue to eject eggs and, incredibly, the host will stand aside to enable the cuckoo to accomplish its task, looking on calmly while the parasite completes the destruction of the host's own reproductive success. Sometimes, one of the ejected host young gets caught on the vegetation just outside the nest rim, but the host parent takes no action while its offspring struggles and slowly dies before its eyes. Hosts will even take time off to mob an adult cuckoo near their nest, before returning to feed a baby cuckoo inside their nest! There could be no more dramatic demonstrations of the host's reliance on its nest contents to recognise its 'own' young.

13.3 Cuckoo hosts and host-egg mimicry

In Britain, the cuckoo parasitises five main hosts (Table 13.1): the meadow pipit *Anthus pratensis* in moorland, the reed warbler *Acrocephalus scirpaceus* in marshland, the dunnock and robin *Erithacus rubecula* in woodland and farmland, and the pied wagtail *Motacilla alba* in open country. The most extensive study of nests parasitised in woodland and farmland was by Owen (1933), who recorded 509 parasitised nests in the Felsted district, Essex, between 1912 and 1933. Of these, the dunnock was easily the most frequent victim, with 302 parasitised nests.

Using the proportion of parasitised nests recorded by the Nest Record Scheme of the British Trust for Ornithology to estimate cuckoo population size, there are currently about 21,000 female cuckoos laying eggs in Britain each year, with about 48% of the parasitism being of dunnocks (Brooke and Davies 1987).

Soon after hatching the cuckoo chick, still naked and blind, ejects the dunnock's eggs out of the nest, one by one. The dunnock never interferes, even as its reproductive success is destroyed in front of its own eyes.

Because the cuckoo is not a common bird, the frequency of parasitism of even the favourite hosts is not high, on average only 5% or less (Table 13.1).

Detailed observations have shown that individual female cuckoos specialise on one particular host species and lay distinctive, and invariant, eggs which match to varying degrees the eggs of their respective hosts (Jourdain 1925; Chance 1940; Baker 1942; Wyllie 1981). These different strains of cuckoo, sometimes called 'gentes' (singular, gens) would remain distinct if, for example, daughters inherited their egg colour from their mother and came to parasitise the same species of host that reared them, perhaps learning the host characteristics through imprinting. By this process, the match between cuckoo and host egg would be maintained over successive cuckoo generations.

The varying degrees of host-egg mimicry are shown in Fig. 13.1. Michael Brooke and I have measured the colour and darkness of the cuckoo eggs from the various gentes and have shown that there are, indeed, significant differences between them (Fig. 13.2). Meadow pipit-cuckoos lay brownish eggs, matching the pipit's own eggs. Reed warbler-cuckoos lay greenish eggs, mimicking the

Table 13.1. The five main hosts of the cuckoo in Britain.

Host	No. breeding pairs in Britain and Ireland*	No. nest record cards (BTO) †		
		Total	With cuckoo	% nests parasitised
Meadow pipit	3.10^6	5,331	142	2.66
Reed warbler	60,000	6,927	384	5.54
Dunnock	5.10^6	23,352	453	1.94
Pied wagtail	500,000	4,945	21	0.42
Robin	5.10^6	12,917	38	0.29

* From Sharrock (1976). † From Glue and Murray (1984), who summarised the nest record cards submitted to the British Trust for Ornithology in the period 1939–1982.

warbler's greenish eggs. Pied wagtail-cuckoos lay pale, greyish-white eggs, again matching those of their host. The cuckoo eggs laid in robin nests are no different in darkness from those of pied wagtail-cuckoos and they are significantly more variable than those of the other gentes, so it is not certain that robin-cuckoos lay distinct eggs (Brooke and Davies 1988). However, as with the other three gentes above, there is a reasonable match between the cuckoo and the host egg.

By sharp contrast, although the dunnock-cuckoo lays a distinctive egg in terms of darkness (Fig. 13.2), there is clearly no colour mimicry (Fig. 13.1). The pale spotted cuckoo egg contrasts sharply with the uniform blue egg of the host. As Gilbert White (1770) remarked in his *Natural History of Selborne*. (Letter V to Daines Barrington), 'you wonder, with good reason, that the hedge sparrows can be induced at all to sit on the egg of the cuckoo without being scandalized at the vast disproportioned size of the supposititious egg; but the brute creation, I suppose, have very little idea of size, colour or number'.

13.4 Experiments with model eggs to test host discrimination

Baker (1913) was the first to suggest that discrimination by hosts selected for egg mimicry by cuckoos. 'The process of perfect adaptation is attained by the slow but sure elimination by the foster parents of those eggs which contrast most distinctively with their own . . . by this means those strains of cuckoos which lay the most ill-adapted eggs gradually die out, while those that lay eggs most like those of the fosterer are enabled to persist.' (See also Baker 1923, 1942.)

Fig. 13.1 Host eggs, cuckoo eggs and model eggs. The left-hand column shows the eggs of four main host species parasitised by the cuckoo in Britain. From top to bottom these are: (1) meadow pipit, (2) reed warbler, (3) dunnock, (4) pied wagtail. The central column is a typical example of a cuckoo egg laid in the corresponding host nest. The right hand column shows an example of a model egg, painted to mimic the corresponding cuckoo egg, for testing host discrimination. In the case of the dunnock (row 3), where there is no colour mimicry by the cuckoo, the model egg was painted pale blue, the same colour as the host egg, so that dunnock responses to mimetic eggs could also be tested.

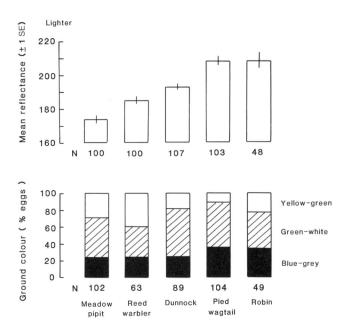

Eggs of different cuckoo gentes

Fig. 13.2 The shade and ground colour of cuckoo eggs from the different gentes in Britain. TOP, the mean (± SE) darkness of cuckoo eggs laid in the nests of the five main hosts. Darkness was measured as reflectance of light (arbitrary units) from a standard source, using a lightmeter with a profiflex attachment. Meadow pipit-cuckoo eggs were significantly darker than reed warbler-cuckoo eggs ($P < 0.01$), which were significantly darker than dunnock-cuckoo eggs ($P < 0.02$), which in turn were significantly darker than pied wagtail-cuckoo eggs ($P < 0.001$) and robin-cuckoo eggs ($P < 0.01$). Robin- and pied wagtail-cuckoo eggs did not differ significantly (*t*-test, 2-tailed). BOTTOM, cuckoo-egg ground colour, divided into three categories. Plate numbers (N) refer to colour plates of the *Methuen Handbook of Colour.* (Kornerup and Wanscher 1967): blue-grey (bluer plates 20–27); green-white (Modal green plate 28); yellow-green (yellower plates 29–5). The differences between cuckoo eggs laid in the five hosts' nests are significant ($\chi^2_8 = 25.3$, $P < 0.005$). Data from Brooke and Davies 1988.

Perhaps because this idea is so eminently reasonable, host discrimination was accepted as the mechanism responsible for egg mimicry by cuckoos and no one bothered to test it by controlled experiments, though Swynnerton (1918) showed that various African birds rejected the eggs of other species and Baker himself (1942) reported that hosts were more likely to desert clutches with eggs laid by the 'wrong' cuckoo gens. Moreover, several authors had commented on the lack of discrimination by dunnocks, which would explain the absence of colour mimicry for this host (Jourdain 1925). For example, Rothschild and Clay (1952) wrote that 'the hedge sparrow will brood almost anything foisted upon it, from cuckoo's eggs to pebbles. It is not surprising to find, therefore, that as no selection takes place, no blue-type egg has been developed by the cuckoo parasitising this species'.

Nevertheless, controlled experiments are needed to test the hypothesis critically, not only because natural ejections of cuckoo eggs may go undetected and hosts may desert clutches for reasons other than parasitism, but also because there are other plausible hypotheses to account for the mimicry (see below). Following a technique pioneered by Stephen Rothstein (1975) for a study of egg discrimination by hosts of a North American brood parasite, the cowbird *Molothrus ater*, Michael Brooke and I used model cuckoo eggs to test discrimination by British cuckoo hosts (Davies and Brooke 1988, 1989*a*). The models, made from gel coat resin, were the same size and weight as real cuckoo eggs and were painted with acrylic paints to match the egg types laid by the various cuckoo gentes (Fig. 13.1). Although cuckoos do not lay a blue mimetic egg for dunnocks, there is a gens of cuckoo in Finland and eastern Europe which specialises on redstarts *Phoenicurus phoenicurus*, and lays a perfect blue egg to match this host egg (Southern 1954; von Haartman 1981). We included a redstart-gens model so we could compare dunnock responses to both mimetic and non-mimetic cuckoo eggs.

We copied the procedure of real cuckoos, parasitising nests one afternoon during the host laying period. Real cuckoos remove one of the host eggs before depositing their own, but our experiments showed that this had no effect on host rejection so our data include cases where we removed a host egg together with those where we simply added the model to the clutch. Most rejections occurred within 3 days of the onset of incubation, many sooner, so we used a 3 day criterion to score acceptance versus rejection.

Table 13.2 shows that meadow pipits, reed warblers and pied wagtails exhibited strong rejection of models which differed from their own eggs in coloration and were most likely to accept a model which mimicked their own eggs, in other words a model which represented the egg laid by their own gens of cuckoo. Models were rejected either by ejection or desertion. Ejection was a clear rejection response but desertion was also primarily a response to a badly matching egg because hosts were more likely to desert in these cases than when there was a mimetic model egg in their nest (Davies and Brooke 1989*a*).

Robins showed a low frequency of rejection of the three model types they were tested with, none of which was a good mimic of their own eggs (Table 13.2). Dunnocks accepted mimetic and non-mimetic models alike. There was only one 'rejection', a desertion of a meadow pipit-cuckoo egg, and this may not have been a response to the model itself.

We wondered if dunnocks had poor colour vision or found it difficult to discern egg colour in their nests, built in dense cover. Perhaps in the dark nest their own blue eggs appeared greyish in colour and so the various models also

Table 13.2. Responses of the five main cuckoo hosts to four types of model cuckoo egg (see Fig. 13.1). The asterisk (*) marks the model type which mimics the host egg (data from Davies and Brooke 1989a).

Host	Meadow pipit-type		Reed warbler-type		Redstart-type		Pied wagtail-type		% rejection (n) of all models unlike host's own eggs (excluding *)	
			% Nests (n) where various types of model cuckoo egg were rejected							
Meadow pipit	22.2*	(27)	26.7	(15)	83.3	(18)	36.0	(25)	48.3	(28/58)
Reed warbler	36.4	(11)	0*	(19)	60.7	(28)	81.3	(16)	61.8	(34/55)
Dunnock	16.7	(6)	0	(4)	0*	(5)	0	(7)	5.9	(1/17)
Pied wagtail	66.7	(18)	71.4	(7)	76.9	(13)	50.0*	(14)	71.0	(27/38)
Robin	18/2	(11)	–		14.3	(7)	28.6	(7)	20.0	(5/25)

There was significant variation in response to the four model types for meadow pipit ($\chi^2_3 = 18.90$, $P < 0.001$) and reed warbler ($\chi^2_3 = 27.05$, $P < 0.001$), but not for the other three hosts.

appeared grey, thus matching in shade. Therefore, we did further experiments with white and black model eggs, which should have been easily detected as different in shade. Black models were accepted at all four nests as were four of five white models (one was deserted).

Both reed warblers (Davies and Brooke 1988) and meadow pipits (Moksnes and Røskaft 1989) are more likely to reject a model cuckoo egg if they have seen a stuffed cuckoo on their nest, but even this treatment did not stimulate dunnocks to reject. At eight nests we placed a stuffed cuckoo on the nest for 10 minutes. The dunnocks approached and showed alarm in all cases, though none attacked the cuckoo as meadow pipits and reed warblers will do (Chance 1940; Duckworth 1991). We then placed a meadow pipit-type model in the nest; it was accepted in all eight cases.

We then worried that the dunnocks might, in fact, be one step ahead of us. Perhaps they realised that the models were simply lumps of resin which would do them no harm, and so accepted them! We tested this by giving dunnocks a whole clutch of model eggs unlike their own. In three out of four nests these were accepted (one was deserted) and the dunnocks then accepted their own eggs back again at the end of the experiment. Other species which show no discrimination (see below) also accepted whole clutches of model eggs, so the birds certainly regarded them as being real (Davies and Brooke 1989*a*).

Reed warblers will reject even mimetic models if these are placed in their nests before they themselves have begun to lay. Very sensibly, they seem to have a rule 'any egg appearing in the nest before I begin to lay cannot be mine', and they eject it. This explains why the cuckoo waits until the host starts laying before parasitising a nest (Davies and Brooke 1988). As a final attempt to induce dunnocks to show rejection, we placed model eggs in completed nests before laying had begun. One out of four redstart-type models was ejected, as was one out of five meadow pipit models (we found them under the nest). These were the only ejections recorded in a total of 48 experiments with model eggs in dunnock nests. They do show, however, that at least rejection is a possibility for dunnocks.

The conclusion is that the degree of egg mimicry by the various gentes of cuckoos (Fig. 13.1) is exactly what we would expect from the degree of discrimination shown by their respective hosts. Meadow pipits, reed warblers and pied wagtails show strong rejection of eggs unlike their own and in all three cases the cuckoo gentes have evolved good mimetic eggs. Robins show less discrimination and the mimicry is less good. Dunnocks show no discrimination, and there is no colour mimicry. Baker's hypothesis is therefore supported experimentally. Other recent experiments have also shown that host discrimination is a selective pressure favouring mimicry by cuckoos (Higuchi 1989).

However, we are left with two problems. First, if there is no host discrimination, why do dunnock-cuckoos still lay a distinctive egg type? Although there is no colour mimicry, the egg not only differs in darkness from the others, it is also no more variable than in the other gentes (Fig. 13.2). What selective pressures might be responsible? Second, why don't dunnocks discriminate when it would clearly pay them to do so? I shall examine these two questions in the following sections.

13.5 Why, if dunnocks do not discriminate, do dunnock-cuckoos lay distinctive eggs?

There are three hypotheses to explain this puzzle.

Selection by secondary hosts

Although there is good evidence for individual specialisation by reed warbler-cuckoos (Wyllie 1981) and meadow pipit-cuckoos (Chance 1940), no detailed study has ever been done on dunnock-cuckoos. If they were less specialised, and frequently laid eggs in other species' nests, then selection by secondary hosts may favour a particular egg type, namely that most likely to be accepted by these other species. Many woodland and farmland passerines have pale mottled eggs and so the dunnock-cuckoo's egg may be a generalised match to a variety of hosts. Even if dunnock-cuckoos turn out to be as specialised as the other gentes, the occasional parasitism of other species (known also from reed warbler-cuckoos and meadow pipit-cuckoos) may be sufficient to select for a particular egg type. A study involving radio-tracking of female 'dunnock-cuckoos', to investigate their degree of host specialisation, would be well worthwhile.

However, host discrimination may not be the only selective pressure influencing the evolution of cuckoo eggs. Two others may be involved.

Predation and cuckoo egg mimicry

Alfred Russel Wallace (1889, p. 216) interpreted bird egg colours as an example of protective coloration and suggested that cuckoo eggs have come to resemble host eggs so the clutch is not made conspicuous to predators: 'if each bird's eggs are to some extent protected by their harmony of colour with their surroundings, the presence of a larger and very differently coloured egg in the nest might be dangerous and lead to the destruction of the whole set. Those cuckoos, therefore, which most frequently placed their eggs among the kinds which they resembled would in the long run leave most progeny, and thus the very frequent accord in colour might have been brought about'. The same suggestion was also made more recently by Harrison (1968).

According to this hypothesis, avian predators, with good colour vision, would presumably select for colour mimicry by cuckoos. But mammalian predators, with no colour vision, could in theory select for shade-matching if a cuckoo egg of different darkness to the dunnock's own eggs made the clutch more conspicuous.

I tested Wallace's hypothesis with the following experiment. At the end of the 1985 breeding season I put quail *Coturnix coturnix japonica* eggs into old nests of blackbirds *Turdus merula* and song thrushes *Turdus philomelos*. The eggs were painted with acrylic paints and three types of clutch were made up, with 21 nests of each type.

(a) Two blue eggs, the same colour and shade as dunnock eggs, plus one grey egg, the same shade as the blue (matched by eye).
(b) Two blue eggs plus one white egg, clearly much paler than the other two.
(c) Two blue eggs plus one very dark brown egg, clearly much darker than the other two.

If Wallace's hypothesis is to explain the distinctive egg type laid by dunnock-cuckoos, then we would predict that clutches containing an egg different in shade—treatments (b) and (c) above—would suffer more predation.

The clutches were indeed found by predators and from the teeth marks on the shells they were mainly mice and grey squirrels. However, there was clearly no difference in predation across the three treatments. After 18 days, 9 of the 21 clutches in (a) had been predated, 9 of the 21 in (b) and 11 of the 21 in (c), and there was no difference in rate of predation up to this time.[1] Two other studies have also shown that clutches with an odd egg do not suffer greater predation (Mason and Rothstein 1987; Davies and Brooke 1988), so there is no support for this as an important selective pressure on cuckoo eggs.

Selection by cuckoos themselves

Another hypothesis stems from the observation that a proportion of parasitised nests are later parasitised again, by a second cuckoo. Cuckoos remove an egg from the nest before laying their own, not because hosts are more likely to reject if there is an extra egg in the nest, but because this increases incubation efficiency of the cuckoo egg and, incidently, provides the laying female with a free meal (Davies and Brooke 1988). Now, it would clearly pay the second cuckoo to pick out the first cuckoo's egg for removal, because there is only room for one cuckoo in the nest; if two hatch out, then one ejects the other and it is the chick from the second laid cuckoo egg which is likely to hatch last and so, being smaller, suffer ejection. Host egg mimicry may evolve, therefore, because it reduces the chance that second cuckoos will be able to discriminate, and remove, a cuckoo egg from the clutch (Davies and Brooke 1988). In our experiments with reed warblers, sometimes a cuckoo came and laid in a nest where we had earlier placed a model egg. There was indeed a tendency for cuckoos to be more likely to remove the model if it was unlike the host eggs in colour, so discrimination by cuckoos themselves may have played a part in the evolution of egg mimicry (Davies and Brooke 1988).

The Brooker family have taken this hypothesis one step further, and in an interesting way which may explain the morphology of the dunnock-cuckoo egg (M.G. and L.C. Brooker 1989; L.C. Brooker *et al.* 1990). They point out that the cuckoo egg could hide from other cuckoos in two ways, either by mimicking the host eggs (as suggested above) or by crypsis in the nest itself. In Australia, the shining bronze-cuckoo *Chrysococcyx lucidus* lays a dark olive-coloured egg in the enclosed and dimly lit nests of thornbills *Acanthiza* spp. The hosts show no egg discrimination, accepting non-mimetic and non-cryptic model eggs alike. It seems possible, therefore, that selection from cuckoos, or indeed any other predator which removed one egg from a clutch, may have played the major role in the evolution of a cryptic cuckoo egg. The Brookers themselves sometimes failed to see the cuckoo egg in the dark thornbill nest and only realised that the nest was parasitised when they felt an extra egg in the clutch!

Could dunnock-cuckoo eggs also be cryptic? Some, at least, look fairly well camouflaged against the nest lining (Fig. 13.3). Further work is needed to discover the frequency of second cuckoos at dunnock nests and to test whether the cuckoo egg coloration is indeed cryptic.

Fig. 13.3 Cuckoo egg in a dunnock nest. Although the brownish spotted cuckoo egg is clearly different from the immaculate blue host eggs, it might be cryptic against the nest lining and so decrease the chance of removal by other cuckoos. Photograph by W.B. Carr.

13.6 Why do dunnocks not discriminate cuckoo eggs?

There are two possible answers to this question (Davies *et al.* 1989).

Acceptance may be better than rejection, for three reasons

(i) In rare cases, hosts may actually gain from raising the parasite's young. For example, parasite giant cowbird *Scaphidura oryzivora* young are raised alongside the host young and the hosts may sometimes benefit from the presence of the parasite because it cleans their own young of botflies (N.G. Smith 1968). It seems impossible that any benefit could accrue to dunnocks, given that the cuckoo chick completely destroys their own reproductive success.

(ii) Defences may be very costly. Some hosts of the cowbird *Molothrus ater* have bills too small to remove the parasite's egg by grasping it and may also be unable to puncture and then eject it because of its unusually thick shell (Spaw and Rohwer 1987). The alternatives are then desertion, to start a new clutch, or acceptance. Acceptance may sometimes be the better option because cowbird hosts raise at least some of their own young from parasitised nests. There is often a seasonal decline in reproductive success so that it could be better to 'make the best of a bad job' and rear a small parasitised brood now rather than an even smaller, unparasitised, brood later. According to this view, by evolving thick-shelled eggs cowbirds have forced many hosts to accept and have thus won the arms race (Rohwer and Spaw 1988). The argument assumes that the hosts are unable to discriminate against the cowbird chick when it hatches (see below). Lisa Petit (1991) has shown that prothonotary warblers *Protonotaria citrea* may fit this hypothesis. They are more likely to reject cowbird eggs (by desertion) when they have alternative nest sites available, which suggests that they may sometimes decide to accept the parasitism as the better of two evils.

However, this hypothesis is unlikely to explain acceptance of cuckoo eggs by dunnocks. First, there is nothing obvious about the dunnock's biology which suggests that rejection or discrimination would be peculiarly costly. They are not small compared to species which reject, nor are their nests different in structure (e.g., depth) or darkness compared to rejector species. Second, the idea of 'adaptive acceptance' is unlikely to apply to cuckoo hosts because, unlike cowbird hosts, they always lose all their own young from a parasitised nest. Desertion must, therefore, always bring a greater benefit than the zero payoff they get from raising the cuckoo chick. Even if there was no chance of another breeding attempt that season, at least it would pay to desert and avoid two months hard work, which must presumably reduce the host's chances of surviving to the next year. Indeed, as we would expect, although small-billed cuckoo hosts do find it more difficult to eject non-mimetic cuckoo eggs from their nests, and so are more likely to reject by desertion, there is no tendency for smaller-billed hosts to reject less (Davies and Brooke 1989*a*). Thus, in cuckoo hosts, rejection costs influence the method of rejection but not rejection frequency.

(iii) If parasitism is rare, it may not be worth investing in counter-adaptation. By analogy, a sudden snowfall causes chaos on British roads, whereas traffic continues to run smoothly in Scandinavia and Canada, where severe conditions are more frequent and so justify investment in expensive snow-clearing equipment. However, this argument does not apply to dunnocks; they suffer the same levels of parasitism as hosts which have evolved discrimination (Table 13.1).

Rejection may be better than acceptance but there is evolutionary lag
According to this view, dunnocks may be behaving maladaptively because they are recent hosts of the cuckoo and simply have not yet had time to evolve counter-adaptations. This presupposes that before a host is exploited by cuckoos it shows, like the dunnock, no rejection of eggs unlike its own.

Michael Brooke and I tested this supposition by placing model eggs in a variety of passerine birds' nests. Some were 'suitable' as cuckoo hosts, namely those with an open nest, accessible to a laying cuckoo, and those which raised their young on an invertebrate diet, necessary for feeding a cuckoo chick. These species showed a range of rejection of eggs unlike their own, from little if any (as in the dunnock) to strong rejection (Fig. 13.4).

If rejection evolves only in response to cuckoos, then we predicted that species 'unsuitable' as hosts would show no rejection, because they would have no history of interaction with the cuckoo. Unsuitable hosts include species that, although their diet is suitable, nest in holes which are inaccessible to cuckoos, and those which, though their nests are accessible, feed their young on seeds. Our prediction was strongly supported; eight of the nine unsuitable host species we tested showed little, if any, rejection of eggs unlike their own (Fig. 13.4).

Some comparisons between closely related species were particularly revealing. Of the four finches we tested (family Fringillidae), only the one that feeds its young predominantly on invertebrates and is therefore suitable as a cuckoo host (the chaffinch *Fringilla coelebs*) showed strong rejection. The three unsuitable hosts, which feed their young mostly on seeds (the greenfinch *Chloris chloris*, the linnet *Acanthis cannabinna* and the bullfinch *Pyrrhula pyrrhula*) showed no rejection. Of the two flycatchers (family Muscicapidae), the spotted flycatcher *Muscicapa striata*, whose open nests are exploitable by cuckoos, showed strong rejection, whereas the pied flycatcher *Ficedula hypoleuca*, whose hole nests are inaccessible to cuckoos, showed no rejection at all. These data indicate that rejection is not related to the taxonomic group to which a species belongs, but rather to the species' evolutionary experience with cuckoos. Moksnes *et al.* (1991) have obtained similar results from experiments in Norway.

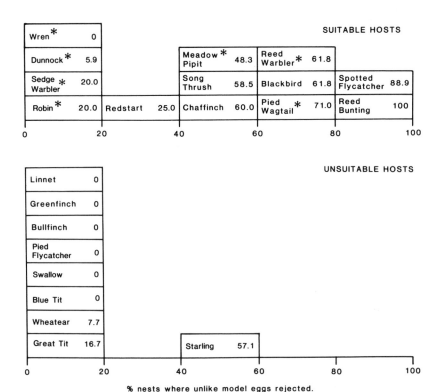

Fig. 13.4 Percentage of nests where a model egg unlike the host's own eggs were rejected. Species suitable as cuckoo hosts (TOP) show varying degrees of rejection, while species that are unsuitable as hosts, and which therefore have no history of co-evolution with the cuckoo, show largely no rejection (BOTTOM). Within suitable species, some rarely used hosts show stronger rejection than the most commonly used hosts (*). From Davies and Brooke 1989*a*.

This comparative study suggests that before cuckoos parasitised meadow pipits, pied wagtails, and other current favourite hosts, these species showed no rejection of eggs unlike their own. Michael Brooke and I could not, of course, go back in time to test this but we could do the next best experiment. The cuckoo breeds from western Europe to Japan but not in Iceland, where it is only a rare vagrant and has never been known to breed. Iceland does, however, have isolated populations of meadow pipits and white wagtails (of which the pied wagtail is a subspecies). We therefore took our model eggs to Iceland. The Icelandic populations bred at low densities and we had to work very hard to find nests but the results were exciting. Both the pipits and wagtails showed much less discrimination against eggs unlike their own than did members of the parasitised populations of these two species in Britain. They did, however, reject some of the pure blue egg models (redstart-type) and so, unlike the unsuitable host species in Fig. 13.4, they were not completely naive. Possibly the Icelandic populations were derived from parasitised populations from other parts of Europe and still have a legacy of their ancestors' egg discrimination, a ghost of adaptation past (Davies and Brooke 1989*a*).

13.7 Are dunnocks recent hosts?

We can conclude that dunnocks do indeed display the characteristics of a species unexploited by cuckoos. Could they be recent hosts? At first, it seems that evidence from the literature would argue against this view. We know from the Gilbert White quote that the dunnock has been a host for over 200 years. We can go back further, to 380 years ago, when, in Shakespeare's *King Lear* (Act I, Scene IV), written in about 1605, the Fool warns Lear that his daughters will prove to be his ruin if he continues to dote on them, just as 'the hedge sparrow fed the cuckoo so long that it had it head bit off by it young'. Shakespeare presumably intended this metaphorically, but there is one curious account where this actually happened! The cuckoo chick normally swallows with its bill closed only once the delivery of the meal from the parent is complete, thus preventing injury to the host which bows deep into its gape. One exception to this usually smooth operation is described by Hens (1949): the cuckoo nestling in question clamped its mouth shut on a hapless dunnock, which suffered fatal injuries!

There is an even more ancient reference to dunnocks and cuckoos in Chaucer's poem *The Parlement of Foules* (1382), where the merlin chastises the cuckoo (line 612): 'Thow mordrer of the heysugge on the braunche that broghte thee forth! Thou rewtheless glotoun!' Heysugge is Old English for the hedge sparrow, so we know that the dunnock has been a common host of the cuckoo for at least 600 years.

Even so, calculations show that the dunnock may, on an evolutionary time-scale, still be a recent host. With a parasitism rate of only 2%, most dunnocks will never encounter cuckoos, so their lack of discrimination will bring only a small selective disadvantage compared with hypothetical 'discriminator' dunnocks. Discrimination would, in fact, take several thousand generations to spread through the dunnock population (Kelly 1987; Davies and Brooke 1989*b*). Most of Britain was covered in primeval woodland until a few thousand years ago (see Chapter 2) and the dunnock is not common in extensive woodland. For example, in the primeval temperate forest of the Bialowieza National Park, eastern Poland, there are on average only 5 pairs per km^2 (Tomialojc *et al.* 1984) and in the New Forest, southern England, 0.6 pairs per km^2 (Glue 1973). This compares with an average of 28 pairs per km^2 on farmland in the midlands of Britain (Williamson 1967). It is possible, therefore, that the dunnock did not become a common victim of the cuckoo until extensive forest clearance occurred 6,500 to 2,500 years ago, within the time which calculations suggest it would take discrimination to spread.

This view must be expressed with some caution, however, because forest clearance may also have led to an increase in moorland species, such as the meadow pipit, which has had time to evolve discrimination and to select for a mimetic egg (Perrins 1983). The only way we could convincingly test the 'evolutionary lag' hypothesis would be to come back in a few thousand years and repeat the model egg experiments. Both Baker (1942) and Lack (1963) claimed that dunnock-cuckoo eggs in some collections had a blue-grey tint to the background and that selection was, therefore, already taking place. The eggs examined by Michael Brooke and I do not convince us of this, and our results with the model eggs would also seem to argue against it. Thus, dunnocks either have no suitable genetic variation that permits rejection, on which selection could act, or rejection is still at a very low frequency in the population (Rothstein 1990).

If rejection, in response to low selective pressure from cuckoos, takes a long time to evolve then this raises the possibility that the varying degrees of rejection shown by suitable cuckoo hosts (Fig. 13.4) represent different stages of a continuing arms race with cuckoos. It is intriguing that some rarely used hosts (unstarred in Fig. 13.4) are just as discriminating against eggs unlike their own, or even more so, than the current favourite hosts (marked by a star). Perhaps they were old favourites of the cuckoo and bear the scars of an arms race their ancestors ran long ago. Possibly they evolved such strong rejection that they drove their cuckoo gentes to extinction, or into the nests of other species as yet unhardened by the struggle (Davies and Brooke 1989*b*, 1991; Moksnes *et al.* 1991).

This must remain a conjecture, but there is good evidence that cuckoos do change their use of hosts with time as shown by recent observations of great spotted cuckoos *Clamator glandarius* in Spain (Arias de Reyna and Hidalgo 1982) and *Cuculus canorus* in Japan (Yamagishi and Fujioka 1986; H. Nakamura 1990), both of which have recently begun to parasitise a new host, the azure-winged magpie *Cyanopica cyana*. In Spain, this new host shows much less rejection of eggs unlike its own than does the common magpie *Pica pica*, a regular and presumably much older host (Arias de Reyna and Hidalgo 1982). These observations support the view that evolutionary lag is a likely possibility for some current hosts.

What might the 'evolution of rejection' entail? It seems unlikely that there is a difference in perceptual abilities between species which reject versus accept non-mimetic eggs. Presumably what evolves in response to parasitism is the decision to reject. However, species which reject foreign eggs (those of con-specifics or brood parasites) do not simply follow a rule 'reject the odd egg', but rather recognise their own eggs and reject foreign eggs whether they are in the minority or form the majority of the clutch (Victoria 1972; Rothstein 1982).

By contrast, species which show no rejection, like the dunnocks, accept even whole clutches of eggs unlike their own (Rothstein 1982; Davies and Brooke 1989a). This suggests that what evolves in response to brood parasitism is not just the decision to reject eggs, but also the ability to recognise one's own eggs, probably by learning (Rothstein 1978). The peculiarity about the dunnock, compared with the other major cuckoo hosts in Britain, may be that it never learns what its own eggs look like.

13.8 Why do dunnocks not discriminate cuckoo chicks?

The dunnock is not alone in its total acceptance of cuckoo chicks. Even hosts which show strong discrimination at the egg stage will accept a parasite chick which differs markedly from their own. For example, the newly hatched cuckoo's pink body and vivid orange gape is quite unlike the reed warbler's chick's black skin and yellow gape with tongue spots. One might suppose that the host parents tolerate cuckoo chicks because they have nothing to compare them with; the cuckoo chick sees to that by hatching first and ejecting the host eggs. However, even when hosts are given a simultaneous choice between their own young and a cuckoo chick, by strapping two nests side by side, they still fail to discriminate (Davies and Brooke 1988).

Richard Dawkins and John Krebs (1979) have suggested that whereas the cuckoo relies on deception to get its egg accepted, it relies on a different trick, manipulation, at the chick stage. They suggest that the hosts may not be able to resist a cuckoo chick any more 'than the junkie can resist his fix'. However, reed warblers will accept a dunnock chick or a reed bunting *Emberiza schoeniculus* chick in among their own brood, raising it as their own, and likewise dunnocks will accept a chaffinch chick (Davies and Brooke 1989a). All these foreign chicks look different from the host's own young. This shows that the cuckoo chick does not possess any special 'drug-like' stimuli necessary to elicit feeding. Rather, it seems as if the hosts simply fail to show any discrimination at the chick stage and will feed any begging mouth in their nest.

Admittedly, chick discrimination may be a more difficult task than egg discrimination. After all, chicks change in appearance daily as they grow and to spot a stranger given such variability in one's own young may present problems. Nevertheless, there are simple rules which dunnocks and reed warblers could use to discriminate. In both species, their young have tongue spots, whereas cuckoo chicks do not. A rule 'only feed chicks with tongue spots' would give both hosts immediate protection against cuckoos.

If no hosts of brood parasites showed rejection of non-mimetic young, we could claim that there may be some design constraint which prevents chick discrimination from evolving in birds. However, in some cases where the brood parasite is raised alongside the host's own young, there is good chick mimicry,

As the young cuckoo grows it may burst out of the host nest, yet the dunnock continues to bow deep into the enormous gape to feed a nestling four times its own body weight.

apparently selected for by host discrimination against chicks unlike their own. For example, near perfect nestling mimicry occurs between the parasitic screaming cowbird *Molothrus rufoaxillaris* and its host the bay-winged cowbird *M. badius*, which is unlikely to be simply attributable to common descent (Lack 1968, p. 94), and nestlings of parasitic viduines mimic the intricate mouth patterns of their particular species of estrildid finch host, which usually do not feed nestlings with mismatching mouth markings (Nicolai 1964).

Rothstein (1990) points out that these examples of nestling mimicry occur in cases where high parasitism rates put very high selection pressures on the host. This may have facilitated the appearance of a host counter-adaptation, chick discrimination, which is hard to evolve. The blind acceptance of a cuckoo chick by hosts of *Cuculus canorus* may, therefore, be explained by the low selective advantage (parasitism rare) together with the higher costs of discriminating chicks than discriminating eggs.

13.9 Summary

The dunnock is a favourite host of the cuckoo in Britain, with about 2% of nests being parasitised. Individual female cuckoos specialise on one host species. Experiments with variously coloured model cuckoo eggs show that the degree of host-egg mimicry exhibited by the different cuckoo gentes reflects the degree of egg discrimination shown by their respective hosts. Unlike other gentes, dunnock-cuckoos do not lay a mimetic egg, as expected from the fact that, in contrast to other hosts, dunnocks show no egg discrimination.

Nevertheless, dunnock-cuckoos still lay a distinctive egg, different in shade from the other cuckoo gentes. Experiments provide no support for predation as an important selective pressure. Either selection by secondary hosts, or by cuckoos themselves (for an egg which is cryptic in the nest) may be involved.

It is unlikely that dunnocks accept non-mimetic eggs because rejection is peculiarly costly for them or of less benefit than for other hosts. Experimental parasitism of species which have no history of interaction with cuckoos shows that before parasitism occurs hosts exhibit no rejection of eggs unlike their own. Dunnocks may, therefore, be recent victims of the cuckoo, lagging behind in their counter-adaptations to a new selective pressure.

Dunnocks, like other hosts of *Cuculus canorus*, show no rejection of cuckoo chicks. The cuckoo chick does not have any special stimuli necessary to manipulate hosts, because hosts accept chicks of other species too, even among broods of their own young.

STATISTICAL ANALYSIS

1. Kolmogorov-Smirnov two-sample test, $K_D = 6$; NS.

14

Beyond dunnocks: sexual conflict, parental care and mating systems

14.1 Three main conclusions

The three main conclusions of the book are as follows. First, the variable mating system of the dunnock reflects the various outcomes of conflicts of interest among males and females, with each individual selected to maximise its own reproductive success, often at the expense of others. Second, conflicts of interest underlie even seemingly cooperative ventures, with no visible squabbles, such as the provisioning of a brood of young by a male-female pair or a two male-female trio. Third, individuals use simple rules to guide their behaviour, rules which on average promote their own reproductive success but which are far from perfect, sometimes leading to maladaptive outcomes such as the provisioning of a brood where a male has no paternity or even the provisioning of young of the wrong species. In this final chapter I shall consider the general interest of these three themes, particularly for studies of other bird species.

14.2 Sexual conflict and mating systems

By 'sexual conflict' I do not mean to imply that there is a conflict between the two classes, male and female, in the sense that one *sex* can win in an evolutionary sense. In sexually reproducing species every individual has one mother and one father and at equilibrium sex ratios the average reproductive success of males and females must be the same. The conflicts of interest which have dominated the dunnock story are conflicts among *individuals*. I have used the term sexual conflict to highlight the fact that there are not only conflicting interests within the same sex, namely competition among males for females and competition among females for males, but also conflicting interests between how a male and a female best maximise their individual reproductive success.

How widespread are such conflicts of interest between males and females? At first sight they might seem to be unusual because the predominant mating system in birds (90% species) is monogamy, with the male-female pair cooperating to raise the young together (Lack 1968). Indeed, David Lack warned that the 'attention inevitably given to unusual pairing habits . . . may mislead the reader as to their prevalence', just as 'if a bird could read our daily newspapers, it might get an exaggerated idea of the frequency of divorce or of glamorous screen stars in our midst' (op. cit. p. 148). He went on to conclude that natural selection usually favoured the 'dull conventional habits of monogamy'. Are dunnocks, then, simply the screen stars of the bird world? Or do the conflicts of interest which dominate their lives also seethe away unnoticed in monogamous species? If so, why don't they, like dunnocks, exhibit variable mating systems?

Sexual conflict over polygyny

Lack suggested that monogamy predominated in birds because 'each male and each female will, on average, leave most descendants if they share in raising a brood' (op. cit. p. 161). This hypothesis certainly seems to explain the obligate monogamy of many seabirds and shorebirds, where male and female share incubation and chick feeding and where the death or removal of one partner leads to complete breeding failure (Oring 1982).

Lack also used his hypothesis to explain the predominance of monogamy in passerine birds. He extended John Crook's (1964) pioneering comparative study of weaver birds (Ploceidae) to show that whereas about a quarter of seed-eating and frugivorous subfamilies are polygynous, monogamy is the rule for every insectivorous passerine subfamily (Lack 1968). The explanation proposed was that while seeds and fruit are sometimes easily exploited, enabling females to raise a brood alone and thus freeing the male to desert and gain extra mates, insects are hard to find, so two parents are needed to raise the brood.

However, recent removal experiments have shown that although male help certainly increases reproductive success in species where both sexes commonly rear the young in a monogamous mating system, it is not essential, especially when food is abundant (Greenlaw and Post 1985; Lyon *et al.* 1987; Wolf *et al.* 1988; Bart and Tornes 1989). Male removal often causes success to decrease to around 50% of that of pair-fed broods, so if a male could gain more than two females he would do better with polygyny and no care than monogamy with care. Even if success was reduced to a half or less and a male could gain only two females, polygyny would still pay a male provided he helped provision at least one of the broods. As predicted, monogamous male passerines readily desert to gain extra mates if given the chance, for example by removal of neighbouring males (J.N.M. Smith *et al.* 1982). Occasional polygyny has been reported in 39% of 122 well-studied European passerines (Møller 1986), and the frequency of this mating system in dunnocks is not unusual by comparison.

Thus, the male-female conflict over polygyny in dunnocks is probably widespread in passerines. The predominance of monogamy in many bird species must arise not, as Lack proposed, because each sex has greatest success with monogamy, but because of the limited opportunities for polygyny. The two constraints which occur in dunnocks are likely to apply to other species too, namely: (a) strong competition among males, which makes it difficult for a male to gain another female, and (b) the aggression shown by females to other females, which decreases the chance that their partner gains another mate and that they suffer, as a consequence, from reduced help with chick rearing (see, for example, Arcese 1989).

Although in many species these constraints restrict most males to monogamy, males may gain increased reproductive success through 'extra pair' copulations and so fertilise some eggs laid by neighbouring females as well as the eggs of their mate (e.g., Westneat 1987; Sherman and Morton 1988; T.R. Birkhead *et al.* 1990). Polygynous males may also increase their success in this way; Gibbs *et al.* (1990) found that male red winged blackbirds *Agelaius phoeniceus* gain on average 20% of their reproductive success from extra-bond fertilisations. There is thus a continuum of varying degrees of monopolisation of females by males from monogamy, through monogamy plus extra-pair copulations, to polygyny.

Sexual conflict over polyandry

Although the dunnock is not unusual in exhibiting monogamy with occasional polygyny, it certainly is unusual (and unique, at least among British passerines) in adding polyandry and polygynandry to its repertoire as well. Polyandry is common, and even when a male enjoys access to several females he rarely has exclusive access but usually shares them with another male in polygynandry (Chapter 3). These two systems can be viewed simply as extensions of extra-bond copulation behaviour with parental care by the additional male. Thus, it is a small step from an unpaired male trespassing next door to gain matings with a monogamous or polygynous female to his remaining on the territory to help feed the young when they hatch, producing polyandry and polygynandry respectively. Alternatively, we can imagine a transition from a monogamous male gaining extra-pair fertilisations with a neighbouring female to his helping with chick feeding in a polygynandrous system (see Chapter 4). Therefore, the key problems to explain are first, why in dunnocks are second males so often able to settle as permanent residents on a territory and second, why do they help in parental care rather than simply steal matings?

Why do male dunnocks have difficulty in maintaining exclusive access to females? Female dunnocks do not have particularly large territories, so the male's problem cannot be that of having to defend an unusually large area. For example, in the Garden female robins *Erithacus rubecula*, which are usually monogamous, had a mean territory of 5,700m² (range 2,550–8,300m², $n = 34$; Harper 1984), which is not significantly different from that of female dunnocks (Chapter 4). M.E. Birkhead (1981) suggested that it is the male-biased sex ratio that gives rise to the high degree of polyandry in dunnocks, but although this certainly affects the frequency of polyandry (Chapter 3) it cannot be the main factor because passerines characterised by a monogamous mating system often have equally male-biased sex ratios (Breitwisch 1989). For example, for robins in the Garden the percentage of males in the population at the start of the breeding season was 54% in 1981, 56% in 1982 and 58% in 1983 (Harper 1984). These figures are similar to the dunnock sex ratios in these three years (from Table 3.2; 54%, 59% and 58% males, respectively). In most passerines, like robins, the 'extra' males remain on bachelor territories, unpaired; the oddity in the dunnock is that most end up as beta males in either polyandry or polygynandry.

What, then, is special about dunnocks? I agree with Barbara and David Snow (1982) that the difficulty a male has in maintaining an exclusive territory is linked to the species' habit of feeding quietly and unobtrusively in dense cover. This makes it easy for intruding males to trespass for long periods undetected and to hide away with the female (Chapter 6). Males of species with more active feeding techniques, or which feed out in the open, would not be able to remain undetected on a territory for so long, unless they stopped feeding. The second ecological factor of importance may be the dunnock's specialisation on very small prey. In Chapter 8 I suggested that this meant that the rate of food delivery to the nest was limited not primarily by food abundance, but rather by the work force available to collect it. Thus, it may pay a female dunnock, more so than species which exploit easily collected bonanzas of large prey (e.g., caterpillars) to have extra male help. The way they gain this is by soliciting matings from beta males as well as alpha males. From the beta male's point of view, his additional help increases the success of the brood and thus improves the survival of the young which were sired by himself.

These two factors, dense cover and small prey, must combine to influence the occurrence of polyandry and polygynandry. Presumably it would pay females of all species to have additional male help. However, if the benefit was not large (two male help not much better than one) and the cost of gaining matings from a second male was high (difficult to escape mate-guarding), then this may be sufficient to tip the balance against females attempting to gain a second male, and against second males joining in the provisioning of the brood. The best way to test these ideas would be to use the comparative approach to see whether species with similar ecology to the dunnock have the same variability of mating system.

14.3 From conflict in polyandry to cooperation

In the dunnocks, polyandry was advantageous to females because more young were raised with more male help, but the increased production of a trio-fed brood did not compensate the alpha male for shared paternity (Chapter 9).

Polyandry occurred, therefore, despite the alpha male's attempts to prevent it. The cooperation in chick feeding arose by default, whenever alpha males were unable to prevent beta males from mating or to evict them from the territory altogether.

We can imagine, however, conditions under which males as well as females would benefit from cooperative polyandry and so the sexual conflict seen in dunnocks would largely disappear. There are two situations in which the benefits of cooperation to males are likely to outweigh the costs of sharing paternity.

(a) Food scarcity and the need for increased parental care

When food is abundant, females may be able to provision a brood successfully alone. However, as food becomes scarcer male help may become essential for success (e.g., Lyon *et al.* 1987; Whittingham 1989; Bart and Tornes 1989). As food becomes yet scarcer, then even a pair may be unable to raise the whole brood and extra helpers may have a marked effect on nestling growth and survival, as seen in some cooperative breeders (Brown 1987; Emlen and Wrege 1991). Under such harsh conditions it may pay a male to share a female in polyandry because the increased production of a trio-fed brood offsets the costs of shared paternity (Gowaty 1981). Figure 14.1 shows how sexual conflict over the acceptance of a second male disappears as the environment becomes harsher.

The best evidence for cooperation arising in this way in birds comes from Uli Reyer's (1980) study of pied kingfishers *Ceryle rudis*. When fish are abundant, a male-female pair can raise a brood alone and males chase off other, unrelated, males which offer to help with provisioning. However, when food is scarce the help of these second males is accepted and it has a considerable effect on fledging success. It is thought that these second males offer to feed the young as a 'payment' to gain access to the female for future breeding attempts and so it only pays the resident male to accept this cost when there is a benefit of extra help in terms of increased production of the current brood. Although this example does not involve polyandry, the costs and benefits are similar to the ones envisaged in the graphical model in Fig. 14.1.

An example which does involve cooperative polyandry comes from a neotropical primate, the saddle-backed tamarin *Saguinus fuscicollis* which, like the dunnock, has a variable mating system including polyandry, monogamy and occasional polygyny (Terborgh and Goldizen 1985; Goldizen 1987, 1989). In polyandry, the males are probably unrelated: both copulate with the female and both help to look after the infants. An increased number of carers seems to be important for successful reproduction because lone pairs rarely attempt to raise young unless they have previous offspring, which act as helpers. Therefore, in a monogamous pair without helpers it probably pays both male and female to allow a second male to join them to gain his help with parental care. In striking contrast to the dunnocks, polyandrous males are rarely aggressive to each other and rarely disrupt each others' copulations. Further data on reproductive success

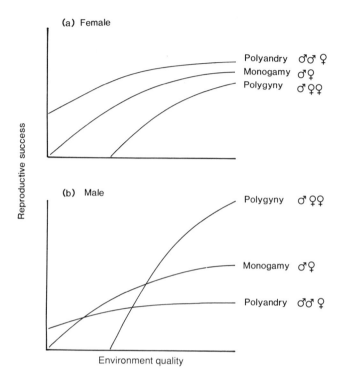

Fig. 14.1 A graphical illustration of sexual conflict. (a) A female's reproductive success per brood is greatest with polyandry, where she has the full-time help of two males, less in monogamy, where she has full-time help of one male, and least in polygyny, where she has only part-time help of one male. For each mating system, success increases with environmental quality. The differences between the curves are greatest in poor environments, where increased male help is of greater benefit. In very rich environments, the curves converge as resources become super-abundant in relation to a brood's needs. (b) A male's reproductive success, calculated from the curves in (a). The monogamy curve is the same as in (a). The polygyny curve is 2× the female polygyny curve because a male's success is the sum of the two females in polygyny. The polyandry curve is half the female polyandry curve, assuming that the two males share paternity equally. In rich environments, male success follows the order polygyny > monogamy > polyandry, the exact reverse of the order for female success; such sexual conflict occurs in dunnocks. In poor environments, however, male success (like female success) may be greatest with polyandry and so there may be less conflict. From Davies 1991.

are needed, but the situation may well be represented by the left hand part of Fig. 14.1, with no male-female conflict over the settlement of a second male. African wild dogs *Lycaon pictus* may provide another example where co-operative polyandry could sometimes be more profitable than monogamy for both males and females. Pairs alone have little if any success at rearing pups and reproductive success increases with more male help (Frame *et al.* 1979; Malcolm and Marten 1982; Moehlman 1988).

It is intriguing to see that in our own species too, although polyandry is a rare marriage system, it is most likely to occur in harsh environments where hard work is needed to maintain an estate (Alexander 1974; J.H. Crook 1980). In the mountain deserts of Ladakh, the Tibetan peasants are polyandrous, with brothers (usually two, but up to five) sharing a wife. In this case, the difficulties of raising a family under harsh ecological circumstances are exacerbated by heavy taxation imposed by landlords. J.H. Crook and S.J. Crook (1988) show that polyandrous families raise more children than monogamous ones, so polyandry clearly increases female reproductive success. In some circumstances, it may also result in two cooperating brothers enjoying greater fitness than either would have achieved in monogamy. However, when economic conditions improve calculations suggest that younger brothers, who are likely to gain less paternity, would do better with monogamy. As predicted, when resources increase, younger brothers often leave and establish a monogamous household and among rich, aristocratic, families even polygyny is common. Although it may not always be justified to apply simple-minded ideas from behavioural ecology to human societies, in this case the parallels with animal mating systems are striking (Borgerhoff Mulder 1991).

At this point it is worth remembering that most of the accentors are montane birds, presumably also inhabiting a harsh environment (Chapter 2). In Japanese accentors *Prunella rubida* (Matuzaki 1991) and alpine accentors *P. collaris* (Dyrcz 1977; M. Nakamura 1990), females often copulate with more than one male and several males sometimes help to feed the young. It would be interesting to study these and the other accentor species, as well as dunnocks living under harsher conditions than lowland farmland and suburban gardens, to assess male and female reproductive success in the various mating systems. Maybe the ancestral state for the behaviour now seen in dunnocks was true cooperative polyandry with males agreeing to share a mate, as represented by the left hand side of Fig. 14.1. When the dunnock colonised lowland regions, where conditions were more benign, and so moved to the right along the *x*-axis in Fig. 14.1, the threshold may have been crossed where it no longer paid a male to agree to share paternity, so producing the sexual conflict which has dominated this book.

(b) Increased competition for territories or mates

A second situation in which it may pay males to cooperate is when there is intense competition for valuable resources. For example, larger coalitions of male lions *Panthera leo* are both more likely to gain control of prides of females and can maintain longer tenure, with the result that reproductive success per individual male increases with coalition size (Packer *et al.* 1988). Males usually cooperate with relatives (brothers, half-brothers), but cooperation also occurs between unrelated individuals and is likewise beneficial to their reproductive success.

Similar situations occur in birds where one male may have difficulty in defending a territory alone because of high intruder pressure, and so joins forces with other males to take over or defend territories. Cooperation may occur between related males (e.g., acorn woodpeckers; Koenig and Mumme 1987) or unrelated males (e.g., pukeko *Porphyrio porphyrio*, Craig 1984; Galapagos hawks, Faaborg and Bednarz 1990). In the acorn woodpeckers and Galapagos hawks, individual males gain greater lifetime reproductive success by co-operating in a group, mainly because of increased survival, though the reasons for the increased survivorship are not clear.

Although in these examples discussed above both males and females may maximise their reproductive success in cooperative polyandry, conflicts of interest are still likely to occur over paternity sharing and parental investment, just as in the dunnocks. Dominant males are expected to attempt to gain more than their fair share of matings and females may attain greatest male help with a mating share that none of the males would choose (Section 12.8). The degree to which dominant males can monopolise the matings without loosing the advantage of cooperation from the subordinate males will depend on two factors (Emlen 1982; Vehrencamp 1983): (i) the alternative options available to the subordinates, who are expected to leave the group when they can do better elsewhere, and (ii) relatedness between the males. In cases where the males are unrelated the dominant male may have to allow the subordinates to gain some chance of paternity to secure their cooperation. When the males are related, however, subordinates gain indirect fitness benefits (*sensu* Brown 1987) from helping to raise the dominant's offspring, so the dominant may be able to impose a greater paternity skew in his own favour. For example, Koenig (1990) showed that subordinate male acorn woodpeckers would still sometimes help to feed the offspring even when they were removed temporarily during the mating period so that they were unable to father any of them. Incidentally, this provides a nice contrast to the dunnock (Chapters 7 and 12) where beta males demand some chance of paternity if they are to help feed the brood. In the acorn woodpeckers the males are close relatives. In the dunnocks they are not, so there is no indirect fitness benefit for a beta male to help raise an alpha male's young.

The advent of new molecular techniques for assigning parentage, together with controlled experiments, should now enable us to explore the delicate balance of fitness gains which are needed to maintain cooperation in breeding groups, despite individual conflicts of interest. Some authors still think about mating systems in terms of which system will be the 'most adaptive' under particular ecological conditions. Perhaps the main message of this discussion is that because conflicts of interest are rife this is not a sensible question to ask. Rather, mating systems should be viewed as *outcomes* of the decisions made by individuals, each selected to maximise its own success.

14.4 Which sex should compete most intensely for mates?

In most animals, the potential reproductive rate of males far exceeds that of females and males are the most active competitors for mates (Trivers 1972; Clutton-Brock 1991). This is particularly clear in insects and mammals where, in many species, males provide just sperm and no help with parental care. In these cases, females often need just one or a few matings to fertilise all their eggs, so female reproductive success is not limited by access to males, only by access to resources, such as food or nest sites. A male, however, has the potential to father offspring at a faster rate than a female can produce them, so male reproductive success is limited by access to females. Thus, it is males who tend to search for females, initiate courtship and fight for mates, rather than vice versa.

However, where males provide parental care, as in most birds, males themselves become a resource important for female reproductive success. Therefore, not only can males increase their success by gaining more females but, in theory, females can also increase their success by gaining more males to help them. Nevertheless, in birds it is still usually males who are the more competitive sex with polygyny being relatively common and polyandry rare (Oring 1982, 1986). Why should this be so? Two likely explanations are, first, that males have the opportunity to desert earlier, for example while females are completing their clutch, thus leaving their partner in a 'cruel bind' (Trivers 1972) and, second, males may have more to gain from desertion because they can potentially fertilise eggs at a faster rate than a female can lay them.

Can we apply these ideas to the dunnocks to explain why, despite the fact that both males and females gain by increasing their number of mates (Chapter 9), it is the males who compete most intensely for females rather than the reverse? Certainly females compete with each other for breeding territories and attempt to gain matings from beta as well as alpha males (Chapter 6). However, there was no indication that females attempted to increase their territory size to gain additional mates when mating systems first formed (Chapter 4). It was the males who behaved so as to maximise their number of mates, by competing to defend as large a territory as possible to encompass several females, by singing and chasing to exclude rivals, and by moving their territories in relation to female distribution (Chapter 4). Males are also larger than females (Chapter 5), perhaps because larger body size has been favoured in competition for mates.

Figure 14.2 shows how male and female dunnock reproductive success increases with an increase in number of mates. Overall, males seem to gain more than females, which may explain why males are the more competitive sex. However, this is only really clear over the region from no mates to one mate. Only females incubate the eggs and brood the young (only females develop brood patches), so a male left alone with a clutch of fertilised eggs or a brood

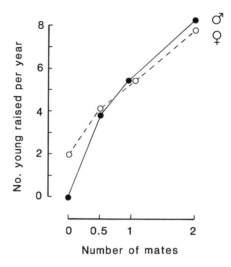

Fig. 14.2 Reproductive success of both male and female dunnocks increases with more mates. Data from Table 9.4 (average of short and long season calculations). For females, 0.5 mates refers to polygyny (one male shared with another female), 1 mate refers to monogamy and 2 mates refers to cooperative polyandry. For males, 0.5 mates refers to cooperative polyandry (assuming paternity shared equally), 1 mate refers to monogamy and 2 mates refers to polygyny. With no mates, males (who do not incubate) have zero success but females can raise some young (Table 11.1; assuming an average of 2.5 broods per year, each with 0.8 young fledged per brood).

of newly hatched young would produce nothing, while a lone female can raise at least part of the brood successfully (Chapter 11). Thus, it is less costly for a male to desert a female than for a female to desert a male.

From one to two mates, however, the increase in male and female success is similar. The limit to the male's increase is set by how females can cope with reduced male help in parental care (Chapters 10 and 11). The limit to the female's increase may be set by the number of eggs that she can cover, and so incubate successfully, and also by the number of young that can squeeze into one nest. Perhaps this explains why, although females increased their clutch size when they anticipated more male help, the increase was, on average, just 0.5 eggs (Chapter 8). A larger increase may not have been possible. The largest broods in the Garden were of five young, which was a tight fit in the nest, while the largest broods commonly found at more northerly latitudes are only of six young (Cramp 1988).

Although Fig. 14.2 may go a little way towards explaining why males are the more competitive sex it cannot be the whole story. First, it only considers the benefits enjoyed from gaining more mates. Additional important factors include the costs of competition and the opportunities to gain mates. Because only females incubate, males are less tied to the vicinity of a nest and so may find it less costly to compete for extra mates and they also have the opportunity to desert first (females cannot desert until the chicks no longer need brooding). Second, the calculations are based on dunnocks as they are now and leave unanswered questions such as why males do not have brood patches or why females do not increase clutch size by laying smaller eggs. Answers to such questions await detailed comparative studies.

14.5 How male and female settlement patterns influence mating systems

Dunnocks: males compete to monopolise female territories

The observations and experiments in Chapter 4 showed that dunnock mating combinations were formed by females first settling in relation to food and nest sites, and males then competing to monopolise the female territories. The number of males associating with a female was thus determined mainly by the outcome of male-male competition. Females later played an active role in influencing which males copulated (Chapter 6) but the initial formation of the mating systems was caused by males distributing over a mosaic of female territories.

The conventional view: female choice of male territories

This settlement pattern appears, at first sight, very different from the conventional view of mating system formation in birds. The conventional view holds that males first settle on breeding territories and females then choose among males and territories, with female choice determining which males gain several mates (polygyny), which gain just one mate (monogamy) and which remain unmated. For example, in summer visitors to temperate regions, the males arrive on the breeding grounds first, set up exclusive territories and advertise for mates. Then the females arrive a week or two later, apparently sampling several male territories before settling to breed.

Several studies have shown that female reproductive success can be influenced both by the quality of a male's territory, for example food and nest sites (e.g., Searcy 1979; Orians 1980; Wittenberger 1981; Askenmo 1984), and also by the amount of male help with parental care (e.g., Alatalo *et al.* 1982, Simmons *et al.* 1986). The 'polygyny threshold model' proposes that females should take account of both these factors when choosing a territory and should settle where their expected reproductive success will be greatest (Verner 1964; Verner and Willson 1966; Orians 1969). For example, the model predicts that a female might choose to settle as a second female with an already mated male on a good territory (i.e., choose polygyny) rather than to settle with a bachelor male on a poor territory (i.e., choose monogamy) if the better access to resources offsets the cost of reduced male care. Predicting exactly how females should settle is difficult unless a great deal is known about the choices available and the costs of sharing (Altmann *et al.* 1977; Davies 1989; Searcy and Yasukawa 1989). Nevertheless, the idea that females should assess both territory quality and prospects of male help with chick feeding seems eminently reasonable and variations of 'polygamy threshold models', involving female choice, have been used to explain most mating systems in passerine birds (Wittenberger 1979).

Is the conventional view wrong?

Is the way dunnock mating systems form unusual—or is the conventional view mistaken? The first point to make is that the difference in the order in which the two sexes settle may not be important. In cases where males settle first and females arrive later, the males may be predicting the places females are likely to favour (e.g., good feeding and nesting sites). What looks like female choice among male territories may, in fact, be female choice of habitat, with the male territories already set up and waiting in anticipation. Tomasz Wesolowski (1987) has pointed out that in *Phylloscopus* warblers, where males arrive first on the breeding grounds, females do not respect male territories when settling, and males have to adjust their territory boundaries to defend the areas females choose to occupy. Therefore, even in cases where males are first to set up territories the females may, as in dunnocks, be ignoring male distribution. The key question is: do females take account of males or only habitat when choosing their breeding territory?

This is a difficult question to answer. Ideally, we would need to remove all the males from the population, or even better, experimentally change their distribution while maintaining the same habitat features. If females were settling in relation to male availability then these experimental changes should cause changes in where females settle. If females were simply choosing habitat, then their settlement should be unaffected. The male removal experiments in Chapter 4, suggested that female dunnocks settled independently of male distribution and that the location and size of female breeding territories was affected only by competition with other females for habitat. What about other species?

The best experiments to tackle this question have concerned pied flycatchers *Ficedula hypoleuca* where males defend nest sites (holes in trees or nest boxes) and their immediate vicinity. Experimental manipulation of male density had no effect on the number of females settling in a habitat (Dale and Slagsvold 1990) but experimental manipulation of nest box site or suitability within a habitat did influence female settlement (Alatalo *et al.* 1986; Slagsvold 1986). These results suggest that females settle in relation to resources, not males. Observations of yellow-headed blackbirds *Xanthocephalus xanthocephalus* also suggest that females settle in relation to habitat characteristics (nest sites in marshes), with mating systems determined simply by the outcome of male-male competition to monopolise the females (Lightbody and Weatherhead 1988).

Further experiments are needed with other species, but it seems as if dunnocks may not be unusual in their formation of mating systems by females choosing habitat and males then competing to defend the females. The conventional view of females choosing among male territories may be mistaken.

But surely male mating status should influence female settlement?

We are now left with a puzzle. It certainly makes good sense for females to choose habitat features, such as food availability and nest sites, because these influence their reproductive success. But so does male help with parental care. Surely then, as proposed in the polygyny threshold model, females should also assess prospects of male help when choosing where to breed. In the yellow-headed blackbirds studied by Lightbody and Weatherhead (1988) males did not provide any help with chick rearing, so provided a female gained sperm to fertilise her eggs she could ignore males altogether and remain indifferent to the mating system that emerged. Leaving the outcome to male-male competition was not costly for them. However, in many other species male help is an important determinant of female success. It has been widely assumed, therefore, that females must take account of male mating status when they settle, and avoid mating with already paired males, who would offer reduced help, unless there are compensating benefits, such as good resources.

One of the main messages from the dunnock study is that it is unwise to assume that if something is advantageous then individuals are bound to do it. For example, it would pay males to discriminate their own young in multiply-sired broods (Chapter 7), it would pay females to eject cuckoo eggs and it would pay to reject cuckoo chicks (Chapter 13); but dunnocks do none of these things. So we should not assume that female birds must necessarily assess male pairing status simply because it seems advantageous for them to do so. Indeed, the evidence so far suggests that female settlement is concerned solely with habitat characteristics.

The only experimental test of whether females assess male pairing status when choosing where to breed is with pied flycatchers. Females suffer from pairing with an already mated male because they are left alone to care for the young (males usually help just their first females) and so raise fewer chicks. By erecting nest boxes in careful sequence, Alatalo *et al.* (1990) arranged for neighbouring boxes, less than 100m apart, to be defended by an unmated male and a mated male, whose first female was incubating a clutch in another box 100–300m away. Boxes were put up at random sites, so there was no difference in territory quality between mated and unmated males. In this situation, females could clearly sample both nest sites and both males (some were seen to do so) yet they showed equal preference for the two options. Furthermore, because of less help with parental care, the females who chose the mated males raised fewer young than females who later chose the unmated males they had rejected. This clever experiment shows that females did not discriminate between mated and unmated males even when they had a simultaneous choice between them, and even though it would clearly have paid them to do so.

Alatalo *et al.* (1981, 1990) suggested that the case of the pied flycatcher might be unusual because males are 'polyterritorial', defending nest boxes some distance apart. Perhaps females are simply unable to assess male pairing status and so are deceived into mating polygynously. All the female sees is a male advertising by a nest box; how is she to know whether he has another female several hundred metres away?

However, the assumption that females ever assess male pairing status may be wrong. Perhaps the dunnock pattern of females competing only for habitat, and males competing for females, will turn out to be widespread in passerine birds. Maybe females follow the rule 'choose suitable habitat and let the outcome of male-male competition determine the mating system'. This rule might be the best one if habitat has an important influence on female reproductive success and if prospects of male help are difficult to assess. Given that it might be costly to sample males (Alatalo *et al.* 1988*a*; Slagsvold *et al.* 1988), and given the problems of predicting the best mating option unless a great deal is known about the costs of sharing and choices available (Davies 1989), a rule based on habitat quality only may be the most practical one for females to use.

Certainly, it is difficult for female dunnocks to predict the mating combination that will emerge, given the fact that they have conflicting interests with males and that males will move if better options become available. Perhaps the only realistic strategy for a female is to ignore male distribution initially and then to try to copulate with all the resident males that defend her to maximise the potential help with chick feeding (Chapter 7).

This discussion highlights the need to understand how animals make decisions. If females do not sample male territories and do not assess male mating status, then models based on costs and benefits of these two parameters are unlikely to predict female settlement. The lesson to be learnt, perhaps, is that we should put as much effort into our bird watching, to see how individuals behave, as we do into detailed measurements of factors influencing reproductive success.

14.6 Summary

Conflicts of interest among males and females are likely to be widespread in birds. In many species, monogamy predominates not because it pays both male and female to breed as a pair but because of limited opportunities to gain additional mates. Sexual conflict may be more likely to lead to a variable mating system in dunnocks than in other species because: (a) their preference for dense cover makes it difficult for a male to monopolise a female; and (b) their specialisation on small prey makes extra male help particularly valuable in chick rearing.

Conflict between male and female over the acceptance of a second male may disappear when environmental conditions are harsh, so that the increased production from cooperative polyandry offsets the costs of shared paternity, and when males gain from cooperation because of intense competition for females. Nevertheless, even in these cases there will be conflicts of interest over paternity sharing and parental investment.

Both male and female dunnocks increase their reproductive success by gaining extra mates, but male reproductive success rises more steeply, which may partly explain why males are the more competitive sex.

The conventional view of mating system formation in birds occurring by females choosing among male territories and assessing male mating status may be mistaken. The dunnock pattern of females competing for habitat and males competing to monopolise females may be widespread.

14.7 Epilogue

The Reverend Morris's charming description of the dunnock, with which this book began, was published three years before the appearance of Darwin's *On the Origin of Species* in 1859. Ten years later, at the age of 59, Morris still rejected Darwin's ideas (Hull *et al.* 1978), so he may have been dismayed by my evolutionary interpretations of dunnock behaviour and would not have agreed that natural selection was responsible for the fit between their curious mating habits and individual reproductive success. However, I hope he would at least have been amused to see that the conflicts of interest and imperfections, which no doubt occurred among his parishioners, also dominate the lives of this little brown bird.

The dunnock's dull appearance and skulking habits may never thrill the observer to the same extent as more colourful species, or those with less retiring behaviour, and its preference for suburban gardens cannot provide the drama of studying birds in wild places. But as a subject for a bird watcher interested in behaviour and mating systems, in my view Morris's 'unobtrusive, sober and unpretending' dunnock is hard to beat.

References

Aichhorn, A. 1969. Lautäusserungen des Schneefinken (*Montifringilla nivalis* Linnaeus) und Begattungsverhalten der Alpenbraunelle (*Prunella collaris* Scopoli). *Verh. Deutschen Zool. Ges.* 32, 690–706.

Alatalo, R.V., Carlson, A., Lundberg, A. and Ulfstrand S. 1981. The conflict between male polygamy and female monogamy: the case of the pied flycatcher, *Ficedula hypoleuca. Am. Nat.* 117, 738–753.

Alatalo, R.V., Lundberg, A. and Ståhlbrandt, K. 1982. Why do pied flycatcher females mate with already-mated males? *Anim. Behav.* 30, 585–593.

Alatalo, R.V., Glynn, C. and Lundberg, A. 1986. Female pied flycatchers choose territory quality not male characteristics. *Nature* 323, 152–153.

Alatalo, R.V., Carlson, A. and Lundberg, A. 1988a. The search cost in mate choice of the pied flycatcher. *Anim. Behav.* 36, 289–291.

Alatalo, R.V., Gottlander, K. and Lundberg, A. 1988b. Conflict or cooperation between parents feeding nestlings in the pied flycatcher *Ficedula hypoleuca* (Pallas). *Ornis Scand.* 9, 31–34.

Alatalo, R.V., Lundberg, A. and Rätti, O. 1990. Male polyterritoriality and imperfect female choice in the pied flycatcher *Ficedula hypoleuca. Behav. Ecol.* 1, 171–177.

Alexander, R.D. 1974. The evolution of social behaviour. *Ann. Rev. Ecol. Syst.* 5, 323–383.

Altmann, S.A., Wagner, S.S. and Lenington, S. 1977. Two models for the evolution of polygyny. *Behav. Ecol. Sociobiol.* 2, 397–410.

Arcese, P. 1989. Intrasexual competition and the mating system in primarily monogamous birds: the case of the song sparrow. *Anim. Behav.* 38, 96–111.

Arias de Reyna, L. and Hidalgo, S.J. 1982. An investigation into egg acceptance by azure-winged magpies and host-recognition of great spotted cuckoo chicks. *Anim. Behav.* 30, 819–823.

Askenmo, C.E.H. 1984. Polygyny and nest site selection in the pied flycatcher. *Anim. Behav.* 32, 972–980.

Baker, E.C.S. 1913. The evolution of adaptation in parasitic cuckoos' eggs. *Ibis* 1913, 384–398.

Baker, E.C.S. 1923. Cuckoo eggs and evolution. *Proc. Zool. Soc. Lond.* 1923, 277–294.

Baker, E.C.S. 1942. *Cuckoo Problems*. London: Witherby.

Barnes, J.A.G. 1970. *Natural History of the Lake District*. Warne.

Bart, J. and Tornes, A. 1989. Importance of monogamous male birds in determining reproductive success: evidence for house wrens and a review of male-removal studies. *Behav. Ecol. Sociobiol.* 24, 109–116.

Bateson, P. 1982. Preferences for cousins in Japanese quail. *Nature* 295, 236–237.

Bateson, P. 1983. Optimal outbreeding. In: *Mate Choice* (ed. P. Bateson), pp. 257–277. Cambridge University Press.

Baylis, J.R. 1981. The evolution of parental care in fishes, with reference to Darwin's rule of male sexual selection. *Env. Biol. Fish.* 6, 223–251.

Beecher, M.D. 1982. Signature systems and kin recognition. *Am. Zool.* 22, 477–490.

Beecher, M.D. 1988. Kin recognition in birds. *Behav. Genet.* 18, 465–482.

Beecher, M.D. and Beecher, I.M. 1979. Sociobiology of bank swallows: reproductive strategy of the male. *Science* 205, 1282–1285.

Beecher, M.D., Beecher, I.M. and Lumpkin, S. 1981a. Parent-offspring recognition in bank swallows. I. Natural history. *Anim. Behav.* 29, 86–94.

Beecher, M.D., Beecher, I.M. and Hahn, S. 1981b. Parent-offspring recognition in bank swallows. II. Development and acoustic basis. *Anim. Behav.* 29, 95–101.

Beissinger, S.R. 1990. Experimental brood manipulations and the monoparental threshold in snail kites. *Am. Nat.* 136, 20–38.

Beletsky, L.D. and Orians, G.H. 1987. Territoriality among male red-winged blackbirds. II. Removal experiments and site dominance. *Behav. Ecol. Sociobiol.* 20, 339–349.

Bertram, B.C.R. 1976. Kin selection in lions and in evolution. In: *Growing Points in Ethology* (ed. P.P.G. Bateson and R.A. Hinde), pp. 281–301. Cambridge University Press.

Bevington, A. 1991. Habitat selection in the dunnock *Prunella modularis* in northern England. *Bird Study* 38, 87–91.

Birkhead, M.E. 1981. The social behaviour of the dunnock *Prunella modularis*. *Ibis* 123, 75–84.

Birkhead, T.R. 1979. Mate guarding in the magpie, *Pica pica*. *Anim. Behav.* 27, 866–874.

Birkhead, T.R. 1982. Timing and duration of mate guarding in magpies, *Pica pica*. *Anim. Behav.* 30, 277–283.

Birkhead, T.R. 1988. Behavioural aspects of sperm competition in birds. *Adv. Study Behav.* 18, 35–72.

Birkhead, T.R. and Møller, A.P. 1991. *Sperm Competition in Birds*. Academic Press, London.

Birkhead, T.R., Pellatt, J.E. and Hunter, F.M. 1988. Extra-pair copulation and sperm competition in the zebra finch. *Nature* 334, 60–62.

Birkhead, T.R., Burke, T., Zann, R., Hunter, F.M. and Krupa, A.P. 1990. Extra-pair paternity and intraspecific brood parasitism in wild zebra finches *Taeniopygia guttata*, revealed by DNA fingerprinting. *Behav. Ecol. Sociobiol.* 27, 315–324.

Birkhead, T.R., Hatchwell, B.J. and Davies, N.B. 1991. Sperm competition and the reproductive organs of the male and female dunnock *Prunella modularis*. *Ibis* 133, 306–311.

Bishton, G. 1985. The diet of nestling dunnocks *Prunella modularis*. *Bird Study* 32, 59–62.

Bishton, G. 1986. The diet and foraging behaviour of the dunnock *Prunella modularis* in a hedgerow habitat. *Ibis* 128, 526–539.

Björklund, M. and Westman, B. 1983. Extra-pair copulations in the pied flycatcher *Ficedula hypoleuca*. *Behav. Ecol. Sociobiol.* 13, 271–275.

Bobr, L.W., Lorenz, F.W. and Ogasawara, F.X. 1964. Distribution of spermatozoa in the oviduct and fertility in domestic birds. I. Residence sites of spermatozoa in fowl oviducts. *J. Reprod. Fert.* 8, 39–47.

Borgerhoff Mulder, M. 1991. Human behavioural ecology. In: *Behavioural Ecology: An Evolutionary Approach*, 3rd edition (ed. J.R. Krebs and N.B. Davies), pp. 69–98. Blackwell Scientific Publications, Oxford.

Breitwisch, R. 1989. Mortality patterns, sex ratios, and parental investment in monogamous birds. *Current Ornithology* 6, 1–50.

Briskie, J.V. in press. Copulation patterns and sperm competition in the polygynandrous Smith's longspur. *Auk*.

Brockmann, H.J., Grafen, A. and Dawkins, R. 1979. Evolutionarily stable nesting strategy in a digger wasp. *J. theor. Biol.* 77, 473–496.

Brooke, M. de L. and Davies, N.B. 1987. Recent changes in host usage by cuckoos *Cuculus canorus* in Britain. *J. Anim. Ecol.* 56, 873–883.

Brooke, M. de L. and Davies, N.B. 1988. Egg mimicry by cuckoos *Cuculus canorus* in relation to discrimination by hosts. *Nature* 335, 630–632.

Brooker, L.C., Brooker, M.G. and Brooker, A.M.H. 1990. An alternative population genetics model for the evolution of egg mimesis and egg crypsis in cuckoos. *J. theor. Biol.* 146, 123–143.

Brooker, M.G. and Brooker, L.C. 1989. The comparative breeding behaviour of two sympatric cuckoos, Horsfield's bronze-cuckoo *Chrysococcyx basalis* and the shining bronze-cuckoo *C. lucidus*, in western Australia: a new model for the evolution of egg morphology and host specificity in avian brood parasites. *Ibis* 131, 528–547.

Brown, J.L. 1987. *Helping and Communal Breeding in Birds*. Princeton University Press.

Brown, J.L., Dow, D.D., Brown, E.R. and Brown, S.D. 1978. Effects of helpers on feeding of nestlings in the grey-crowned babbler *Pomatostomus temporalis*. *Behav. Ecol. Sociobiol.* 4, 43–59.

Bryant, D.M. 1988. Energy expenditure and body mass changes as measures of reproductive costs in birds. *Funct. Ecol.* 2, 23–34.

Burke, T. and Bruford, M.W. 1987. DNA fingerprinting in birds. *Nature* 327, 149–152.

Burke, T., Davies, N.B., Bruford, M.W. and Hatchwell, B.J. 1989. Parental care and mating behaviour of polyandrous dunnocks *Prunella modularis* related to paternity by DNA fingerprinting. *Nature* 338, 249–251.

Byle, P.A.F. 1987. Behaviour and ecology of the dunnock *Prunella modularis*. PhD thesis, University of Cambridge.

Byle, P.A.F. 1990. Brood division and parental care in the period between fledging and independence in the dunnock *Prunella modularis*. *Behaviour* 113, 1–20.

Byle, P.A.F. 1991. Parental provisioning in the dunnock *Prunella modularis*: the effects of a variable mating system. *Ibis* 133, 199–204.

Campbell, B. 1952. *Bird Watching for Beginners*. Penguin, Harmondsworth.

Chance, E.P. 1940. *The Truth about the Cuckoo*. County Life, London.

Charnov, E.L. 1982. *The Theory of Sex Allocation*. Princeton University Press.

Chase, I.D. 1980. Cooperative and non-cooperative behavior in animals. *Am. Nat.* 115, 827–857.

Cheng, K.M., Burns, J.T. and McKinney, F. 1983. Forced copulation in captive mallards. III. Sperm competition. *Auk* 100, 302–310.

Clark, H.H.G. 1930. Extraordinary display by a pair of hedge sparrows. *Brit. Birds* 23, 342.

Clutton-Brock, T.H. (ed.) 1988. *Reproductive Success: Studies of Individual Variation in Contrasting Breeding Systems*. Chicago University Press.

Clutton-Brock, T.H. 1991. *The Evolution of Parental Care*. Princeton University Press.

Clutton-Brock, T.H., Guinness, F.E. and Albon, S.D. 1982. *Red Deer: The Behaviour and Ecology of Two Sexes*. Chicago University Press.

Coulson, J.C. 1966. The influence of pair-bond and age on the breeding biology of the Kittiwake gull *Rissa tridactyla*. *J. Anim. Ecol.* 35, 269–279.

Craig, J.L. 1980. Pair and group breeding behaviour of a common gallinule, the pukeko *Porphyrio porphyrio melanotus*. *Anim. Behav.* 28, 593–603.

Craig, J.L. 1984. Are communal pukeko caught in the Prisoner's dilemma? *Behav. Ecol. Sociobiol.* 14, 147–150.

Cramp, S. (ed.) 1988. *The Birds of the Western Palearctic*, Vol. V. Oxford University Press.

Crook, J.H. 1964. The evolution of social organisation and visual communication in the weaver birds (Ploceinae). *Behaviour* Suppl. 10, 1–178.

Crook, J.H. 1980. Social change in Indian Tibet. *Social Sci. Info.* 19, 139–166.

Crook, J.H. and Crook, S.J. 1988. Tibetan polyandry: problems of adaptation and fitness. In: *Human Reproductive Behaviour: A Darwinian Perspective* (ed. L. Betzig, M. Borgerhoff Mulder and P. Turke), pp. 97–114. Cambridge University Press.

Crook, J.R. and Shields, W.M. 1985. Sexually selected infanticide by adult male barn swallows. *Anim. Behav.* 33, 754–761.

Curio, E. 1983. Why do young birds reproduce less well? *Ibis* 125, 400–404.

Dale, S. and Slagsvold, T. 1990. Random settlement of female pied flycatchers *Ficedula hypoleuca*: significance of male territory size. *Anim. Behav.* 39, 231–243.

Darwin, C. 1859. *On the Origin of Species*. John Murray, London.

Darwin, F. (ed.) 1887. *The Life and Letters of Charles Darwin, Including An Autobiographical Chapter*. John Murray, London.

Davies, N.B. 1978. Territorial defence in the speckled wood butterfly (*Pararge aegeria*): the resident always wins. *Anim. Behav.* 26, 138–147.

Davies, N.B. 1983. Polyandry, cloaca-pecking and sperm competition in dunnocks. *Nature* 302, 334–336.

Davies, N.B. 1985. Cooperation and conflict among dunnocks *Prunella modularis* in a variable mating system. *Anim. Behav.* 33, 628–648.

Davies, N.B. 1986. Reproductive success of dunnocks *Prunella modularis* in a variable mating system. I. Factors influencing provisioning rate, nestling weight and fledging success. *J. Anim. Ecol.* 55, 123–138.

Davies, N.B. 1989. Sexual conflict and the polygamy threshold. *Anim. Behav.* 38, 226–234.

Davies, N.B. 1991. Mating systems. In: *Behavioural Ecology: An Evolutionary Approach*, 3rd edition (ed. J.R. Krebs and N.B. Davies), pp. 263–294. Blackwell Scientific Publications, Oxford.

Davies, N.B. and Brooke, M. de L. 1988. Cuckoos versus reed warblers: adaptations and counteradaptations. *Anim. Behav.* 36, 262–284.

Davies, N.B. and Brooke, M. de L. 1989a. An experimental study of co-evolution between the cuckoo *Cuculus canorus* and its hosts. I. Host egg discrimination. *J. Anim. Ecol.* 58, 207–224.

Davies, N.B. and Brooke, M. de L. 1989b. An experimental study of co-evolution between the cuckoo *Cuculus canorus* and its hosts. II. Host egg markings, chick discrimination and general discussion. *J. Anim. Ecol.* 58, 225–236.

Davies, N.B. and Brooke, M. de L. 1991. Co-evolution of the cuckoo and its hosts. *Sci. Amer.* 264(1), 92–98.

Davies, N.B. and Hatchwell, B.J. 1992. The value of male parental care and its influence on reproductive allocation by male and female dunnocks *Prunella modularis*. *J. Anim. Ecol.* 61.

Davies, N.B. and Houston, A.I. 1986. Reproductive success of dunnocks *Prunella modularis* in a variable mating system. II. Conflicts of interest among breeding adults. *J. Anim. Ecol.* 55, 139–154.

Davies, N.B. and Lundberg, A. 1984. Food distribution and a variable mating system in the dunnock *Prunella modularis*. *J. Anim. Ecol.* 53, 895–912.

Davies, N.B. and Lundberg, A. 1985. The influence of food on time budgets and timing of breeding of the dunnock *Prunella modularis*. *Ibis* 127, 100–110.

Davies, N.B., Bourke, A.F.G. and Brooke, M. de L. 1989. Cuckoos and parasitic ants: interspecific brood parasitism as an evolutionary arms race. *Trends Ecol. Evol.* 4, 274–278.

Davies, N.B., Hatchwell, B.J., Robson, T. and Burke, T. 1992. Paternity and parental effort in dunnocks *Prunella modularis*: how good are male chick-feeding rules? *Anim. Behav.* 43, 729–745.

Dawkins, R. 1976. *The Selfish Gene*. Oxford University Press.

Dawkins, R. and Krebs, J.R. 1979. Arms races between and within species. *Proc. Roy. Soc. Lond. B* 205, 489–511.

Delamain, J. 1929. Extraordinary sexual display by a pair of hedge sparrows. *Brit. Birds* 23, 19–20.

Desrochers, A. In press. Age and foraging success in European blackbirds: variation among and within individuals. *Anim. Behav.*

Dow, D.D. 1977. Reproductive behaviour of the noisy miner, a communally breeding honeyeater. *Living Bird* 16, 163–185.

Downhower, J.F. and Armitage, K.B. 1971. The yellow-bellied marmot and the evolution of polygamy. *Am. Nat.* 105, 355–370.

Duckworth, J.W. 1991. Responses of breeding reed warblers *Acrocephalus scirpaceus* to mounts of sparrowhawks *Accipiter nisus*, cuckoo *Cuculus canorus* and jay *Garrulus glandarius*. *Ibis* 133, 68–74.

Dyrcz, A. 1977. Nest helpers in the alpine accentor, *Prunella collaris*. *Ibis* 119, 215.

Emlen, S.T. 1982. The evolution of helping. II. The role of behavioural conflict. *Am. Nat.* 119, 40–53.

Emlen, S.T. and Oring, L.W. 1977. Ecology, sexual selection, and the evolution of mating systems. *Science* 197, 215–223.

Emlen, S.T. and Wrege, P.H. 1988. The role of kinship in helping decisions among white-fronted bee-eaters. *Behav. Ecol. Sociobiol.* 23, 305–315.

Emlen, S.T. and Wrege, P.H. 1991. Breeding biology of white-fronted bee-eaters at Nakuru: the influence of helpers on breeder fitness. *J. Anim. Ecol.* 60, 309–326.

Faaborg, J. and Bednarz, J.C. 1990. Galapagos and Harris' hawks: divergent causes of sociality in two raptors. In *Cooperative Breeding in Birds*: (ed. P.B. Stacey and W.D. Koenig), pp. 359–383. Cambridge University Press.

Faaborg, J., de Vries, T.J., Patterson, C.B. and Griffin, C.R. 1980. Preliminary observations on the occurrence and evolution of polyandry in the Galapagos hawk, *Buteo galapagoensis*. *Auk* 97, 581–590.

Falla, R.A., Sibson, R.B. and Turbott, E.G. 1970. *A Field Guide to the Birds of New Zealand*, 2nd edition. Collins, London.

Fatio, M.V. 1864. Note sur une particularité de l'appareil reproducteur mâle chez l'*Accentor alpinus*. *Rev. et Mag. de Zool.* 27, 65–67.

Foltz, D.W. and Hoogland, J.L. 1981. Analysis of the mating system in the black-tailed prairie dog *Cynomys ludovicianus* by likelihood of paternity. *J. Mamm.* 62, 706–712.

Frame, L.H., Malcolm, J.R., Frame, G.W. and Lawick, H. van 1979. Social organisation of African wild dogs *Lycaon pictus* on the Serengeti plains, Tanzania 1967–1978. *Z. Tierpsychol.* 50, 225–249.

Fuller, R.J. 1982. *Bird Habitats in Britain.* Poyser, Calton.

Gardam, W.H. 1929. Extraordinary display by a pair of hedge sparrows. *Brit. Birds* 23, 199–200.

Gibbons, D.W. 1987. Juvenile helping in the moorhen *Gallinula chloropus*. *Anim. Behav.* 35, 170–181.

Gibbs, H.L., Weatherhead, P.J., Boag, P.T., White, B.N., Tabak, L.M. and Hoysak, D.J. 1990. Realized reproductive success of polygynous red-winged blackbirds revealed by DNA markers. *Science* 250, 1394–1397.

Glue, D. 1973. The breeding birds of a New Forest valley. *Brit. Birds* 66, 461–472.

Glue, D. and Murray, E. 1984. Cuckoo hosts in Britain. *British Trust for Ornithology News* 134, 5.

Goldizen, A.W. 1987. Facultative polyandry and the role of infant-carrying in wild saddle-backed tamarins, *Sanguinus fuscicollis*. *Behav. Ecol. Sociobiol.* 20, 99–109.

Goldizen, A.W. 1989. Social relationships in a cooperatively polyandrous group of tamarins, *Sanguinus fuscicollis*. *Behav. Ecol. Sociobiol.* 24, 79–89.

Gowaty, P.A. 1981. An extension of the Orians-Verner-Willson model to account for mating systems besides polygyny. *Am. Nat.* 118, 851–859.

Gowaty, P.A. and Karlin, A.A. 1984. Multiple maternity and paternity in apparently monogamous eastern bluebirds (*Sialia sialis*). *Behav. Ecol. Sociobiol.* 15, 91–95.

Grafen, A. 1988. On the uses of data on lifetime reproductive success. In: *Reproductive Success: Studies of Individual Variation in Contrasting Breeding Systems* (ed. T.H. Clutton-Brock). Chicago University Press.

Grant, B.R. and Grant, P.R. 1989. *Evolutionary Dynamics of a Natural Population: The Large Cactus Finch of the Galápagos.* Chicago University Press.

Grant, P.R. 1986. *Ecology and Evolution of Darwin's Finches.* Princeton University Press.

Greenlaw, J.S. and Post, W. 1985. Evolution of monogamy in seaside sparrows *Ammodramus maritimus*: tests of hypotheses. *Anim. Behav.* 33, 373–383.

Greenwood, P.J. and Harvey, P.H. 1982. The natal and breeding dispersal of birds. *Ann. Rev. Ecol. Syst.* 13, 1–21.

Gustafsson, L. and Sutherland, W.J. 1988. The costs of reproduction in the collared flycatcher *Ficedula albicollis*. *Nature* 335, 813–815.

Gyllensten, U.B., Jakobsson, S. and Temrin, H. 1990. No evidence for illegitimate young in monogamous and polygynous warblers. *Nature* 343, 168–170.

Hamilton, W.D. 1964. The genetical evolution of social behaviour. *J. theor. Biol.* 7, 1–52.

Hamilton, W.D. 1979. Wingless and fighting males in fig wasps and other insects. In: *Sexual Selection and Reproductive Competition in Insects* (ed. M.S. Blum and N.A. Blum), pp. 167–220. Academic Press, New York.

Hanken, J. and Sherman, P.W. 1981. Multiple paternity in Belding's ground squirrel litters. *Science* 212, 351–353.

Harper, D.G.C. 1984. Economics of foraging and territoriality in the European robin *Erithacus rubecula*. PhD thesis, Cambridge University.

Harper, D.G.C. 1985. Brood division in robins. *Anim. Behav.* 33, 466–480.

Harrison, C.J.O. 1968. Egg mimicry in British cuckoos. *Bird Study*, 15, 22–28.

Harrison, C.J.O. 1982. *An Atlas of the Birds of the Western Palaearctic.* Collins, London.

Harrison, C.J.O. and Binfield, F.G. 1967. Cloaca-pecking and copulation in the dunnock. *Bird Study* 14, 192–193.

Hartley, H.T. 1930. Extraordinary display by a pair of hedge sparrows. *Brit. Birds* 23, 342.

Hatchwell, B.J. and Davies, N.B. 1990. Provisioning of nestlings by dunnocks *Prunella modularis* in pairs and trios: compensation reactions by males and females. *Behav. Ecol. Sociobiol.* 27, 199–209.

Hatchwell, B.J. and Davies, N.B. 1992a. An experimental study of mating competition in monogamous and polyandrous dunnocks *Prunella modularis*: I. Mate guarding and copulations. *Anim. Behav.* 43, 595–609.

Hatchwell, B.J. and Davies, N.B. 1992b. An experimental study of mating competition in monogamous and polyandrous dunnocks *Prunella modularis*. II. Influence of removal and replacement experiments on mating systems. *Anim. Behav.* 43, 611–622.

Hegner, R.E. and Wingfield, J.C. 1987. Effects of experimental manipulation of testosterone levels on parental investment and breeding success in male house sparrows. *Auk* 104, 462–469.

Hens, P. 1949. Jonge Koekoek *(Cuculus canorus)* doodt een van zijn pleegouders. [Young cuckoo kills one of its fosterers]. *Limosa* 22, 329–330.

Hett, W.S. 1936. *Aristotle: Minor Works. On Marvellous Things Heard.* Heinemann, London.

Higuchi, H. 1989. Responses of the bush warbler *Cettia diphone* to artificial eggs of *Cuculus* cuckoos in Japan. *Ibis* 131, 94–98.

Holmes, W.G. and Sherman, P.W. 1982. The ontogeny of kin recognition in two species of ground squirrels. *Am. Zool.* 22, 491–517.

Houston, A.I. and Davies, N.B. 1985. The evolution of cooperation and life history in the dunnock *Prunella modularis*. In: *Behavioural Ecology: Ecological Consequences of Adaptive Behaviour* (ed. by R.M. Sibly and R.H. Smith), pp. 471–487. Blackwell Scientific Publications, Oxford.

Hrdy, S.B. 1977. *The Langurs of Abu: Female and Male Strategies of Reproduction.* Harvard University Press, Cambridge, Massachusetts.

Hrdy, S.B. 1979. Infanticide among animals: a review, classification, and examination of the implications for the reproductive strategies of females. *Ethol. Sociobiol.* 1, 13–40.

Hull, D.L., Tessner, P.D. and Diamond, A.M. 1978. Planck's principle: do younger scientists accept new scientific ideas with greater alacrity than older scientists? *Science* 202, 717–723.

Jakobsson, S. 1988. Territorial fidelity of willow warbler *(Phylloscopus trochilus)* males and success in competition over territories. *Behav. Ecol. Sociobiol.* 22, 79–84.

Järvinen, O. and Väisänen, R.A. 1978. Long-term population changes of the most abundant south Finnish forest birds during the past 50 years. *J. Orn.* 119, 441–449.

Jeffreys, A.J., Wilson, V. and Thein, S.L. 1985. Hypervariable "minisatellite" regions in human DNA. *Nature* 314, 67–73.

Jenner, E. 1788. Observations on the natural history of the cuckoo. *Phil. Trans. Roy. Soc. Lond.* 78, 219–237.

Jones, G. 1986. Sexual chases in sand martins *Riparia riparia*: cues for males to increase their reproductive success. *Behav. Ecol. Sociobiol.* 19, 179–185.

Jourdain, F.C.R. 1925. A study of parasitism in the cuckoos. *Proc. Zool. Soc. Lond.* 1925, 639–667.

Kacelnik, A. 1984. Central place foraging in starlings *Sturnus vulgaris*. I. Patch residence times. *J. Anim. Ecol.* 53, 283–299.

Karanja, W.K. 1982. The biology of the dunnock *Prunella modularis* with special emphasis on its breeding biology. PhD thesis, Oxford University.

Kelly, C. 1987. A model to explore the rate of spread of mimicry and rejection in hypothetical populations of cuckoos and their hosts. *J. theor. Biol.* 125, 283–299.

Kikkawa, J. 1966. Population distribution of land birds in temperate rainforest of southern New Zealand. *Trans. Roy. Soc. N.Z. Zool.* 7, 215–277.

Kluyver, H.N. 1951. The population ecology of the great tit *Parus m. major* L. *Ardea* 39, 1–135.

Knystautas, A.J. 1987. Siberian accentor. *Brit. Birds* 80, 669–670.

Koenig, W.D. 1990. Opportunity of parentage and nest destruction in polygynandrous acorn woodpeckers *Melanerpes formicivorus*. *Behav. Ecol.* 1, 55–61.

Koenig, W.D. and Mumme, R.L. 1987. *Population Ecology of the Cooperatively Breeding Acorn Woodpecker*. Princeton University Press.

Komdeur, J. 1991. Cooperative breeding in the Seychelles warbler. PhD thesis, University of Cambridge.

Kornerup, A. and Wanscher, J.H. 1967. *Methuen Handbook of Colour*. Methuen, London.

Krebs, J.R. 1982. Territorial defence in the great tit *Parus major*: do residents always win? *Behav. Ecol. Sociobiol.* 11, 185–194.

Krebs, J.R. 1984. Citation classic. *Current Contents* 15 (No. 15), p. 14.

Krebs, J.R., Erichsen, J.T., Webber, M.I. and Charnov, E.L. 1977. Optimal prey-selection by the great tit *Parus major*. *Anim. Behav.* 25, 30–38.

Krebs, J.R., Kacelnik, A. and Taylor, P. 1978. Test of optimal sampling by foraging great tits. *Nature* 275, 27–31.

Labov, J.B. 1980. Factors influencing infanticidal behaviour in wild male house mice *Mus musculus*. *Behav. Ecol. Sociobiol.* 6, 297–303.

Lack, D. 1963. Cuckoo hosts in England. *Bird Study* 10, 185–202.

Lack, D. 1965. *The Life of the Robin*. Witherby, London.

Lack, D. 1966. *Population Studies of Birds*. Clarendon Press, Oxford.

Lack, D. 1968. *Ecological Adaptations for Breeding in Birds*. Methuen, London.

Lake, P.E. 1975. Gamete production and the fertile period with particular reference to domesticated birds. *Symp. Zool. Soc. Lond.* 35, 225–244.

Lifjeld, J.T. and Slagsvold, T. 1990. Manipulations of male parental investment in polygynous pied flycatchers, *Ficedula hypoleuca*. *Behav. Ecol.* 1, 48–54.

Lightbody, J.P. and Weatherhead, P.J. 1988. Female settling patterns and polygyny: tests of a neutral-mate-choice hypothesis. *Am. Nat.* 132, 20–33.

Lyon, B.E., Montgomerie, R.D. and Hamilton, L.D. 1987. Male parental care and monogamy in snow buntings. *Behav. Ecol. Sociobiol.* 20, 377–382.

McDonald, D.B. 1989. Cooperation under sexual selection: age graded changes in a lekking bird. *Am. Nat.* 134, 709–730.

McKinney, F., Derrickson, S.R. and Mineau, P. 1983. Forced copulation in waterfowl. *Behaviour* 86, 250–294.

Magrath, R.D. 1991. Nestling weight and juvenile survival in the blackbird, *Turdus merula*. *J. Anim. Ecol.* 60, 335–351.

Malcolm, J.R. and Marten, K. 1982. Natural selection and the communal rearing of pups in African wild dogs *Lycaon pictus*. *Behav. Ecol. Sociobiol.* 10, 1–13.

Mallory, F.F. and Brooks, R.J. 1978. Infanticide and other reproductive strategies in the collared lemming *Dicrostonyx groenlandicus*. *Nature* 273, 144–146.

Mason, P. and Rothstein, S.I. 1987. Crypsis versus mimicry and the color of shiny cowbird eggs. *Am. Nat.* 130, 161–167.

Matuzaki, Y. 1991. [Breeding biology and mating system of the Japanese hedge sparrow *Prunella rubida* at Mt. Norikura]. Master's thesis, in Japanese. Joetsu University of Education, Japan.

Maynard Smith, J. 1977. Parental investment—a prospective analysis. *Anim. Behav.* 25, 1–9.

Maynard Smith, J. 1982. *Evolution and the Theory of Games*. Cambridge University Press.

Maynard Smith, J. and Ridpath, M.G. 1972. Wife sharing in the Tasmanian native hen, *Tribonyx mortierii*: a case of kin selection? *Am. Nat.* 106, 447–452.

Mead, C.J. and Clark, J.A. 1989. Report on bird-ringing for 1988. *Ringing and Migration* 10, 158–196.

Meiklejohn, A.H. 1930. Extraordinary display by a pair of hedge sparrows. *Brit. Birds* 23, 255.

Moehlman, P.D. 1988. Intraspecific variation in canid social systems. In: *Carnivore Behavior: Ecology and Evolution* (ed. J.L. Gittleman), pp. 143–163. Chapman and Hall, London.

Moksnes, A. and Røskaft, E. 1989. Adaptations of meadow pipits to parasitism by the common cuckoo. *Behav. Ecol. Sociobiol.* 24, 25–30.

Moksnes, A., Røskaft, E., Braa, A.T., Korsnes, L., Lampe, H.M. and Pedersen, H.Ch. 1991. Behavioural responses of potential hosts towards artificial cuckoo eggs and dummies. *Behaviour* 116, 64–89.

Møller, A.P. 1986. Mating systems among European passerines: a review. *Ibis* 128, 234–250.

Møller, A.P. 1987. Mate guarding in the swallow *Hirundo rustica*: an experimental study. *Behav. Ecol. Sociobiol.* 21, 119–123.

Møller, A.P. 1988a. Paternity and parental care in the swallow, *Hirundo rustica*. *Anim. Behav.* 36, 996–1005.

Møller, A.P. 1988b. Testes size, ejaculate quality and sperm competition in birds. *Biol. J. Linn. Soc.* 33, 273–283.

Montagu, G. 1802. *Ornithological Dictionary*. London.

Morris, F.O. 1856. *A History of British Birds*. Groombridge, London.

Morton, E.S. 1987. Variation in mate guarding intensity by male purple martins. *Behaviour* 101, 211–224.

Nakamura, H. 1990. Brood parasitism by the cuckoo *Cuculus canorus* in Japan and the start of new parasitism on the azure-winged magpie *Cyanopica cyana*. *Jap. J. Ornithol.* 39, 1–18.

Nakamura, M. 1990. Cloacal protuberance and copulatory behaviour of the alpine accentor *(Prunella collaris)*. *Auk* 107, 284–295.

Newton, I. 1988. Age and reproduction in the sparrowhawk. In: *Reproductive Success: Studies of Individual Variation in Contrasting Breeding Systems* (ed. T.H. Clutton-Brock), pp. 201–219. Chicago University Press.

Newton, I. (ed.) 1989. *Lifetime Reproduction in Birds*. Academic Press.

Nicolai, J. 1964. Der Brutparasitismus der Viduinae als ethologisches Problem. *Z. Tierpsychol.* 21, 129–204.

Nur, N. 1984. Feeding frequencies of nestling blue tits *Parus caeruleus*: costs, benefits and a model of optimal feeding frequency. *Oecologia* 65, 125–137.

Ollason, J.C. and Dunnet, G.M. 1978. Age, experience and other factors affecting the breeding success of the fulmar, *Fulmarus glacialis*, in Orkney. *J. Anim. Ecol.* 47, 961–976.

Orians, G.H. 1969. On the evolution of mating systems in birds and mammals. *Am. Nat.* 103, 589–603.

Orians, G.H. 1980. *Some Adaptations of Marsh-Nesting Blackbirds*. Princeton University Press.

Orians, G.H. and Pearson, N.E. 1979. On the theory of central place foraging. In: *Analysis of Ecological Systems* (ed. D.J. Horn, R. Mitchell and G.R. Stair), pp. 155–177. Ohio State University Press, Columbus.

Oring, L.W. 1982. Avian mating systems. In: *Avian Biology*, Vol. 6 (ed. D.S. Farner, J.R. King and K.C. Parkes), pp. 1–92. Academic Press, London.

Oring, L.W. 1986. Avian polyandry. In: *Current Ornithology*, Vol. 3 (ed. R.F. Johnston), pp. 309–351. Plenum Press, New York.

Orton, K. 1930. Extraordinary display by a pair of hedge sparrows. *Brit. Birds* 23, 255.

Owen, J.H. 1933. The cuckoo in the Felsted district. *Rep. Felsted School Sci. Soc.* 33, 25–39.

Packer, C. and Pusey, A.E. 1983. Adaptations of female lions to infanticide by incoming males. *Am. Nat.* 121, 716–728.

Packer, C., Herbst, L., Pusey, A.E., Bygott, J.D., Hanby, J.P., Cairns, S.J. and Borgerhoff Mulder, M. 1988. Reproductive success of lions. In: *Reproductive Success* (ed. T.H. Clutton-Brock), pp. 363–383. University of Chicago Press.

Page, R.E., Robinson, G.E. and Fondrk, M.K. 1989. Genetic specialists, kin recognition and nepotism in honey-bee colonies. *Nature* 338, 576–579.

Parker, G.A. 1970. Sperm competition and its evolutionary consequences in the insects. *Bio. Rev.* 45, 525–567.

Parker, G.A. 1979. Sexual selection and sexual conflict. In: *Sexual Selection and Reproductive Competition in Insects* (ed. M.S. Blum and N.A. Blum), pp. 123–166. Academic Press, New York.

Parker, G.A. and Macnair, M.R. 1978. Models of parent-offspring conflict. I. Monogamy. *Anim. Behav.* 26, 97–110.

Patterson, C.B., Erckmann, W.J. and Orians, G.H. 1980. An experimental study of parental investment and polygyny in male blackbirds. *Am. Nat.* 116, 757–769.

Peck, A.L. 1970. *Aristotle: Historia Animalium*, Vol II. Heinemann, London.

Perrins, C.M. 1965. Population fluctuations and clutch size in the great tit *Parus major* L. *J. Anim. Ecol.* 34, 601–647.

Perrins, C.M. 1979. *British Tits*. New Naturalist, Collins.

Perrins, C.M. 1983. The effect of man on the British avifauna. *Proc. Symp. Birds and Man*, 5–29. Johannesburg.

Perrins, C.M. and McCleery, R.H. 1985. The effect of age and pair bond on the breeding success of great tits *Parus major*. *Ibis* 127, 306–315.

Petit, L.J. 1991. Adaptive tolerance of cowbird parasitism by prothonotary warblers: a consequence of nest-site limitation? *Anim. Behav.* 41, 425–432.

Pettifor, R.A., Perrins, C.M. and McCleery, R.H. 1988. Individual optimization of clutch size in great tits. *Nature* 336, 160–162.

Praz, J.-C. 1980. *Prunella modularis*. In: *Verbreitungsatlas der Brutvögel der Schweiz/Atlas des Oiseaux nicheurs de Suisse* (ed. A. Schifferli, P. Géroudet and R. Winkler). Schweizerische Vogelwarte, Sempach/Station ornithologique suisse de Sempach.

Rabenold, P.P., Rabenold, K.N., Piper, W.H., Haydock, J. and Zack, S.W. 1990. Shared paternity revealed by genetic analysis in cooperatively breeding wrens. *Nature* 348, 538–540.

Rackham, O. 1986. *The History of the Countryside*. Dent, London.

Reyer, H.-U. 1980. Flexible helper structure as an ecological adaptation in the pied kingfisher *Ceryle rudis rudis*. *Behav. Ecol. Sociobiol.* 6, 219–227.

Robertson, R.J. and Stutchbury, B.J. 1988. Experimental evidence for sexually selected infanticide in tree swallows. *Anim. Behav.* 36, 749–753.

Rohwer, S. and Spaw, C.D. 1988. Evolutionary lag versus bill-size constraints: a comparative study of the acceptance of cowbird eggs by old hosts. *Evol. Ecol.* 2, 27–36.

Rollin, C.N. 1929. Extraordinary sexual display by a pair of hedge sparrows. *Brit. Birds* 23, 103.

Rothschild, M. and Clay, T. 1952. *Fleas, Flukes and Cuckoos: a Study of Bird Parasites*. Collins, New Naturalist, London.

Rothstein, S.I. 1975. An experimental and teleonomic investigation of avian brood parasitism. *Condor* 77, 250–271.

Rothstein, S.I. 1978. Mechanisms of avian egg recognition: additional evidence for learned components. *Anim. Behav.* 26, 671–677.

Rothstein, S.I. 1982. Successes and failures in avian egg and nestling recognition with comments on the utility of optimality reasoning. *Am. Zool.* 22, 547–560.

Rothstein, S.I. 1990. A model system for co-evolution: avian brood parasitism. *Ann. Rev. Ecol. Syst.* 21, 481–508.

Sanderson, R.F. 1968. Cloaca pecking in the dunnock. *Bird Study* 15, 213.

Sasvari, L. 1986. Reproductive effort of widowed birds. *J. Anim. Ecol.* 55, 553–564.

Schmid-Hempel, P. 1986. Do honeybees get tired? The effect of load weight on patch departure. *Anim. Behav.* 34, 1243–1250.

Schwartz, O.A. and Armitage, K.B. 1980. Genetic variation in social mammals: the marmot model. *Science* 207, 665–667.

Searcy, W.A. 1979. Female choice of mates: a general model for birds and its application to red-winged blackbirds *Agelaius phoeniceus*. *Am. Nat.* 114, 77–100.

Searcy, W.A. and Yasukawa, K. 1989. Alternative models of territorial polygyny in birds. *Am. Nat.* 134, 323–343.

Selous, E. 1933. *Evolution of Habit in Birds*. Constable, London.

Sharrock, J.T.R. 1976. *The Atlas of Breeding Birds in Britain and Ireland*. Poyser, Calton.

Sherman, P.W. 1981. Reproductive competition and infanticide in Belding's ground squirrels and other animals. In: *Natural Selection and Social Behaviour* (ed. R.D. Alexander and D. Tinkle), pp. 311–331. Chiron Press, New York.

Sherman, P.W. and Morton, M.L. 1988. Extra-pair fertilizations in mountain white-crowned sparrows. *Behav. Ecol. Sociobiol.* 22, 413–420.

Shugart, G.W. 1988. Uterovaginal sperm-storage glands in sixteen species with comments on morphological differences. *Auk* 105, 379–385.

Sibley, C.G. 1970. A comparative study of the egg-white proteins of passerine birds. *Bull. Peabody Mus. Nat. Hist.* 32, 131 pp.

Sibley, C.G. and Ahlquist, J.E. 1981. The relationships of the accentors *(Prunella)* as indicated by DNA-DNA hybridization. *J. Orn.* 122, 369–378.

Simmons, R.E., Smith, P.C. and MacWhirter, R.B. 1986. Hierarchies among northern harrier *Circus cyaneus* harems and the costs of polygyny. *J. Anim. Ecol.* 55, 755–771.

Slagsvold, T. 1986. Nest site settlement by the pied flycatcher: does the female choose her mate for the quality of his house or himself? *Ornis Scand.* 17, 210–220.

Slagsvold, T., Lifjeld, J.T., Stenmark, G. and Breiehagen, T. 1988. On the cost of searching for a mate in female pied flycatchers *Ficedula hypoleuca. Anim. Behav.* 36, 433–442.

Smith, J.N.M., Yom-Tov, Y. and Moses, R. 1982. Polygyny, male parental care and sex ratio in song sparrows: an experimental study. *Auk* 99, 555–564.

Smith, N.G. 1968. The advantages of being parasitized. *Nature* 219, 690–694.

Smith, S.M. 1988. Extra-pair copulations in black-capped chickadees: the role of the female. *Behaviour* 107, 15–23.

Snow, B.K. and Snow, D.W. 1982. Territory and social organisation in a population of dunnocks *Prunella modularis. J. Yamashina Inst. Ornithol.* 14, 281–292.

Snow, D.W. 1988. *A Study of Blackbirds*, 2nd edition. British Museum, London.

Snow, D.W. and Snow, B.K. 1983. Territorial song of the dunnock *Prunella modularis. Bird Study* 30, 51–56.

Southern, H.N. 1954. Mimicry in cuckoo's eggs. In: *Evolution as a Process* (ed. J. Huxley, A.C. Hardy and E.B. Ford), pp. 219–232. Allen and Unwin, London.

Spaw, C.D. and Rohwer, S. 1987. A comparative study of eggshell thickness in cowbirds and other passerines. *Condor* 89, 307–318.

Spencer, R. and Hudson, R. 1977. Report on bird-ringing for 1975. *Bird Study* (Supplement) 24, 1–64.

Stacey, P.B. 1979. Kinship, promiscuity, and communal breeding in the acorn woodpecker. *Behav. Ecol. Sociobiol.* 6, 53–66.

Stacey, P.B. and Ligon, J.D. 1987. Territory quality and dispersal options in the acorn woodpecker, and a challenge to the habitat saturation model of cooperative breeding. *Am. Nat.* 130, 654–676.

Stephens, D.W. and Krebs, J.R. 1986. *Foraging Theory.* Princeton University Press.

Svensson, L. 1970. *Identification Guide to European Passerines.* Naturhistoriska Riksmuseet, Stockholm.

Swynnerton, C.F.M. 1918. Rejections by birds of eggs unlike their own: with remarks on some of the cuckoo problems. *Ibis* 1918, 127–154.

ten Cate, C. and Bateson, P. 1989. Sexual imprinting and a preference for "supernormal" partners in Japanese quail. *Anim. Behav.* 38, 356–358.

Terborgh, J. and Goldizen, A.W. 1985. On the mating system of the cooperatively breeding saddle-backed tamarin, *Sanguinus fuscicollis. Behav. Ecol. Sociobiol.* 16, 293–299.

Tinbergen, J.M. and Boerlijst, M.C. 1990. Nestling weight and survival in individual great tits *Parus major. J. Anim. Ecol.* 59, 1113–1127.

Tinbergen, J.M. and Daan, S. 1990. Family planning in the great tit *Parus major*: optimal clutch size as integration of parent and offspring fitness. *Behaviour* 114, 161–190.

Tinbergen, N. 1953. *The Herring Gull's World.* Collins, New Naturalist, London.

Tinbergen, N. 1972. *The Animal in its World*, Vol. 1, *Field Studies*. George Allen and Unwin, London.

Tinbergen, N. 1974. *Curious Naturalists*. Penguin, Harmondsworth.

Tomek, T. 1980. Nesting of dunnock *Prunella modularis*. *Acta Zool. Cracov.* 24(13), 539–560.

Tomek, T. 1988. The breeding biology of the dunnock *Prunella modularis* in the Ojcow National Park (south Poland). *Acta Zool. Cracov.* 31, 115–166.

Tomialojc, L., Wesolowski, T. and Walankiewicz, W. 1984. Breeding bird community of a primaeval temperate forest (Bialowieza National Park, Poland). *Acta Orn. Warszawa* 20(3), 241–310.

Trivers, R.L. 1972. Parental investment and sexual selection. In: *Sexual Selection and the Descent of Man* (ed. B. Campbell), pp. 136–179. Aldine, Chicago.

Trivers, R.L. 1974. Parent-offspring conflict. *Am. Zool.* 14, 249–264.

Tuomenpuro, J. 1989. Habitat preferences and territory size of the dunnock *Prunella modularis* in southern Finland. *Ornis Fenn.* 66, 133–141.

Tuomenpuro, J. 1990. Population increase and breeding biology of the dunnock *Prunella modularis* in southern Finland. *Ornis Fenn.* 67, 33–44.

Vaurie, C. 1959. *The Birds of the Palearctic Fauna: Order Passeriformes*. H.F. and G. Witherby, London.

Vehrencamp, S.L. 1983. A model for the evolution of despotic versus egalitarian societies. *Anim. Behav.* 31, 667–682.

Veiga, J.P. 1990*a*. Infanticide by male and female house sparrows. *Anim. Behav.* 39, 496–502.

Veiga, J.P. 1990*b*. Sexual conflict in the house sparrow: interference between polygynously mated females versus asymmetric male investment. *Behav. Ecol. Sociobiol.* 27, 345–350.

Verma, O.P. and Cherms, F.L. 1965. The appearance of sperm and their persistency in storage tubules of turkey hens after a single insemination. *Poult. Sci.* 44, 609–613.

Verner, J. 1964. Evolution of polygamy in the long-billed marsh wren. *Evolution* 18, 252–261.

Verner, J. and Willson, M.F. 1966. The influence of habitats on mating systems of North American passerine birds. *Ecology* 47, 143–147.

Victoria, J.K. 1972. Clutch characteristics and egg discriminative ability of the African village weaverbird, *Ploceus cucullatus*. *Ibis* 114, 367–376.

von Haartman, L. 1981. Co-evolution of the cuckoo *Cuculus canorus* and a regular host. *Ornis Fenn.* 58, 1–10.

Waldman, B. 1988. The ecology of kin recognition. *Ann. Rev. Ecol. Syst.* 19, 543–571.

Waldman, B. and Bateson, P. 1989. Kin association in Japanese quail chicks. *Ethology* 80, 283–291.

Wallace, A.R. 1889. *Darwinism: An Exposition of the Theory of Natural Selection with Some of its Applications*. Macmillan, London.

Walters, S.M. 1981. *The Shaping of Cambridge Botany*. Cambridge University Press.

Wesolowski, T. 1987. Polygyny in three temperate forest passerines (with a critical reevaluation of hypotheses for the evolution of polygyny). *Acta Orn.* 23, 273–302.

Westneat, D.F. 1987. Extra-pair fertilizations in a predominantly monogamous bird: genetic evidence. *Anim. Behav.* 35, 877–886.

Westneat, D.F. 1988. Male parental care and extrapair copulations in the indigo bunting. *Auk* 105, 149–160.

Westneat, D.F. 1990. Genetic parentage in the indigo bunting: a study using DNA fingerprinting. *Behav. Ecol. Sociobiol.* 27, 67−76.

Wetton, J.H., Carter, R.E., Parkin, D.T. and Walters, D. 1987. Demographic study of a wild house sparrow population by DNA "fingerprinting". *Nature* 327, 147−149.

White, G. 1770. Letters to Daines Barrington, Letter IV. In: *The Natural History of Selborne* (ed. R. Mabey) 1977, Penguin, Harmondsworth.

Whittingham, L.A. 1989. An experimental study of paternal behaviour in red-winged blackbirds. *Behav. Ecol. Sociobiol.* 25, 73−80.

Whittingham, L.A., Taylor, P.D. and Robertson, R.J. in press. Confidence of paternity and male parental care. *Am. Nat.*

Wiley, R.H., Hatchwell, B.J. and Davies, N.B. 1991. Recognition of individual males' songs by female dunnocks: a mechanism increasing the number of copulatory partners and reproductive success. *Ethology* 88, 145−153.

Williams, G.C. 1966a. *Adaptation and Natural Selection.* Princeton University Press.

Williams, G.C. 1966b. Natural selection, the costs of reproduction, and a refinement of Lack's principle. *Am. Nat.* 100, 687−690.

Williamson, K. 1967. The bird community of farmland. *Bird Study* 14, 210−226.

Wingfield, J.C. 1984. Androgens and mating systems: testosterone-induced polygyny in normally monogamous birds. *Auk* 101, 665−671.

Winkler, D.W. 1987. A general model for parental care. *Am. Nat.* 130, 526−543.

Wittenberger, J.F. 1979. The evolution of mating systems in birds and mammals. In: *Handbook of Behavioral Neurobiology*, Vol. 3 (ed. P. Marler and J. Vandenbergh), pp. 271−349. Plenum Press, New York.

Wittenberger, J.F. 1981. *Animal Social Behaviour.* Duxbury Press, Boston.

Wolf, L., Ketterson, E.D. and Nolan, V. Jr. 1988. Paternal influence on growth and survival of dark-eyed junco young: do parental males benefit? *Anim. Behav.* 36, 1601−1618.

Woolfenden, G.E. and Fitzpatrick, J.W. 1984. *The Florida Scrub Jay: Demography of a Cooperative Breeding Bird.* Princeton University Press.

Wooller, R.D., Bradley, J.S., Skira, I.J. and Serventy, D.L. 1990. Reproductive success of short-tailed shearwaters *Puffinus tenuirostris* in relation to their age and breeding experience. *J. Anim. Ecol.* 59, 161−170.

Wrege, P.H. and Emlen, S.T. 1987. Biochemical determination of parental uncertainty in white-fronted bee-eaters. *Behav. Ecol. Sociobiol.* 20, 153−160.

Wright, J. and Cuthill, I. 1989. Manipulation of sex differences in parental care. *Behav. Ecol. Sociobiol.* 25, 171−181.

Wyllie, I. 1981. *The Cuckoo.* Batsford, London.

Yamagishi, S. and Fujioka, M. 1986. Heavy brood parasitism by the common cuckoo *Cuculus canorus* on the azure-winged magpie *Cyanopica cyana*. *Tori* 34, 91−96.

Author Index

Ahlquist, J.E. 12
Aichhorn, A. 114
Alatalo, R.V. 5, 170, 245–8
Albon, S.D. 7
Alexander, R.D. 241
Altmann, S.A. 246
Arcese, P. 237
Arias de Reyna, L. 231
Aristotle, 215
Armitage, K.B. 5, 120
Askenmo, C.E.H. 245

Baker, E.C.S. 217–8, 220, 223, 231
Barnes, J.A.G. 11
Bart, J. 236, 239
Bateson, P. 129
Baylis, J.R. 5
Bednarz, J.C. 242
Beecher, I.M. 88, 128
Beecher, M.D. 88, 128, 130
Beissinger, S.R. 184
Beletsky, L.D. 82
Bertram, B.C.R. 143
Bevington, A. 11
Binfield, F.G. 107
Birkhead, M.E. 5–6, 34, 42–3, 237
Birkhead, T.R. 88, 98, 109, 111–3, 121, 204, 237
Bishton, G. 16, 132
Björklund, M. 100
Boag, P.T. 121, 237
Bobr, L.W. 114
Boerlijst, M.C. 138
Borgerhoff Mulder, M. 241
Bourke, A.F.G. 227
Braa, A.T. 228, 231
Bradley, J.S. 78
Breiehagen, T. 248
Breitwisch, R. 238
Briskie, J.V. 96
Brockmann, H.J. 157
Brooke, M. de L. 216–8, 220–3, 225–32
Brooker, A.M.H. 226

Brooker, L.C. 226
Brooker, M.G. 226
Brooks, R.J. 128, 143
Brown, E.R. 170
Brown, J.L. 38, 170, 239, 242
Brown, S.D. 170
Bruford, M.W. 121–6
Bryant, D.M. 171
Burke, T. 121–6, 196–8, 200–1, 203, 206–8, 237
Burns, J.T. 105
Bygott, J.D. 241
Byle, P.A.F. 69–70, 127, 173, 177–8

Cairns, S.J. 241
Campbell, B. 5
Carlson, A. 5, 248
Carter, R.E. 121
Chance, E.P. 217, 223–4
Charnov, E.L. 7, 196
Chase, I.D. 170
Chaucer, G. 230
Cheng, K.M. 105
Cherms, F.L. 114
Clark, H.H.G. 107
Clark, J.A. 29
Clay, T. 220–1
Clutton-Brock, T.H. 7, 154, 243
Coulson, J.C. 78
Craig, J.L. 69, 96, 242
Cramp, S. 10, 12, 14–15, 244
Crook, J.H. 236, 241
Crook, J.R. 143
Crook, S.J. 241
Curio, E. 78
Cuthill, I. 170, 173

Daan, S. 4
Dale, S. 246
Darwin, C. 15, 249
Darwin, F. 15

Dawkins, R. 7, 157, 232
de Vries, T.J. 155
Delamain, J. 107
Derrickson, S.R. 105
Desrochers, A. 79
Diamond, A.M. 249
Dow, D.D. 96, 170
Downhower, J.F. 5
Duckworth, J.W. 223
Dunnet, G.M. 78
Dyrcz, A. 241

Emlen, S.T. 5–6, 120, 193, 239, 242
Erckmann, W.J. 184
Erichsen, J.T. 196

Faaborg, J. 155, 242
Falla, R.A. 14
Fatio, M.V. 114
Fitzpatrick, J.W. 4, 103
Foltz, D.W. 120
Fondrk, M.K. 129
Frame, G.W. 241
Frame, L.H. 241
Fujioka, M. 231
Fuller, R.J. 11

Gardam, W.H. 107
Gibbons, D.W. 170
Gibbs, H.L. 121, 237
Glue, D. 218, 230
Glynn, C. 246
Goldizen, A.W. 239
Gottlander, K. 170
Gowaty, P.A. 120, 239
Grafen, A. 7, 157
Grant, B.R. 4
Grant, P.R. 4, 7
Greenlaw, J.S. 236
Greenwood, P.J. 33
Griffin, C.R. 155
Guinness, F.E. 7
Gustafsson, L. 4, 138, 149
Gyllensten, U.B. 121

Hahn, S. 128
Hamilton, L.D. 170, 236, 239
Hamilton, W.D. 7, 38
Hanby, J.P. 241
Hanken, J. 120
Harper, D.G.C. 177, 237–8
Harrison, C.J.O. 12, 14, 107, 224
Hartley, H.T. 107

Harvey, P.H. 33
Hatchwell, B.J. 18, 68–71, 82–4, 97–102,
    104–6, 109, 113, 121–6, 135, 144,
    149–51, 172, 174–6, 181–5, 188–91,
    193, 196–203, 206–8, 211
Haydock, J. 121
Hegner, R.E. 195
Hens, P. 230
Herbst, L. 241
Hett, W.S. 215
Hidalgo, S.J. 231
Higuchi, H. 223
Holmes, W.G. 129
Hoogland, J.L. 120
Houston, A.I. 135, 154, 156–7, 159–61,
    164, 170, 192, 210
Hoysak, D.J. 121, 237
Hrdy, S.B. 142–3
Hudson, R. 29
Hull, D.L. 249
Hunter, F.M. 121, 204, 237

Jakobsson, S. 82, 121
Järvinen, O. 14
Jeffreys, A.J. 120–1
Jenner, E. 215–6
Jones, G. 97
Jourdain, F.C.R. 217, 220

Kacelnik, A. 134, 196
Karanja, W.K. 5–6, 42–3
Karlin, A.A. 120
Kelly, C. 230
Ketterson, E.D. 236
Kikkawa, J. 14
Kluyver, H.N. 4
Knystautas, A.J. 13
Koenig, W.D. 4, 38, 144, 155, 242
Komdeur, J. 155
Kornerup, A. 220
Korsnes, L. 228, 231
Krebs, J.R. 4–5, 82, 84, 196, 232
Krupa, A.P. 121, 237

Labov, J.B. 129, 143
Lack, D. 4–6, 193, 231, 233, 236–7
Lake, P.E. 96
Lampe, H.M. 228, 231
Lawick, H. van 241
Lenington, S. 246
Lessells, C.M. 157
Lifjeld, J.T. 184, 248
Lightbody J.P. 246–7
Ligon, J.D. 155

Lorenz, F.W. 114
Lumpkin, S. 128
Lundberg, A. 5, 51, 60–4, 66–8, 88, 170, 245–8
Lyon, B.E. 170, 236, 239

McCleery, R.H. 78, 149
McDonald, D.B. 19
McKinney, F. 105
Macnair, M.R. 192
MacWhirter, R.B. 245
Magrath, R.D. 79, 138
Malcolm, J.R. 241
Mallory, F.F. 128, 143
Marten, K. 241
Mason, P. 225
Matuzaki, Y. 241
Maynard Smith, J. 5–7, 38
Mead, C.J. 29
Meiklejohn, A.H. 107
Mineau, P. 105
Moehlman, P.D. 241
Moksnes, A. 223, 228, 231
Møller, A.P. 100, 106, 111–2, 210, 236
Montagu, G. 11
Montgomerie, R.D. 170, 236, 239
Morris, F.O. 1–2, 249
Morton, E.S. 210
Morton, M.L. 120, 237
Moses, R. 236
Mumme, R.L. 4, 38, 155, 242
Murray, E. 218

Nakamura, H. 231
Nakamura, M. 114, 241
Newton, I. 78, 154
Nice, M.M. 6
Nicolai, J. 233
Nolan, V. Jr. 236
Nur, N. 169, 171

Ogasawara, F.X. 114
Ollason, J.C. 78
Orians, G.H. 48, 82, 134, 184, 245
Oring, L.W. 5–6, 236, 243
Orton, K. 107
Owen, J.H. 216

Packer, C. 143, 241
Page, R.E. 129
Parker, G.A. 7, 88, 192
Parkin, D.T. 121
Patterson, C.B. 155, 184

Pearson, N.E. 134
Peck, A.L. 215
Pedersen, H.Ch. 228, 231
Pellatt, J.E. 204
Perrins, C.M. 4, 78, 138, 149, 155, 231
Petit, L.J. 227
Pettifor, R.A. 149
Piper, W.H. 121
Post, W. 236
Praz, J.-C. 11
Pusey, A.E. 143, 241

Rabenold, K.N. 121
Rabenold, P.P. 121
Rackham, O. 11
Rätti, O. 247–8
Reyer, H.-U. 239
Ridpath, M.G. 38
Robertson, R.J. 144, 210
Robinson, G.E. 129
Robson, T. 196–8, 200–1, 203, 206–8
Rohwer, S. 227
Rollin, C.N. 107
Røskaft, E. 223, 228, 231
Rothschild, M. 220–1
Rothstein, S.I. 221, 225, 231–3

Sanderson, R.F. 107
Sasvari, L. 170
Schmid-Hempel, P. 196
Schwartz, O.A. 120
Searcy, W.A. 245–6
Selous, E. 106–7
Serventy, D.L. 78
Shakespeare, W. 230
Sharrock, J.T.R. 218
Sherman, P.W. 120, 129, 142, 237
Shields, W.M. 143
Shugart, G.W. 112
Sibley, C.G. 12
Sibson, R.B. 14
Simmons, R.E. 245
Skira, I.J. 78
Slagsvold, T. 184, 246, 248
Smith, J.N.M. 236
Smith, N.G. 227
Smith, P.C. 245
Smith, S.M. 106
Snow, B.K. 5–6, 10, 42–3, 101, 238
Snow, D.W. 5–6, 10, 42–3, 101, 103, 238
Southern, H.N. 221
Sozou, P. 192
Spaw, C.D. 227
Spencer, R. 29
Stacey, P.B. 96, 155

Ståhlbrandt, K. 245
Stenmark, G. 248
Stephens, D.W. 196
Stutchbury, B.J. 144
Sutherland, W.J. 4, 138, 149
Svensson, L. 74
Swynnerton, C.F.M. 220

Tabak, L.M. 121, 237
Taylor, P.D. 196, 210
Temrin, H. 121
ten Cate, C. 129
Terborgh, J. 239
Tessner, P.D. 249
Thein, S.L. 120
Tinbergen, J.M. 4, 138
Tinbergen, N. 4, 6, 21
Tomek, T. 87, 132
Tomialojc, L. 230
Tornes, A. 236, 239
Trivers, R.L. 5, 7, 176, 243
Tuomenpuro, J. 11, 14, 41, 42–3
Turbott, E.G. 14

Ulfstrand, S. 5, 248

Väisänen, R.A. 14
Vaurie, C. 12
Vehrencamp, S.L. 242
Veiga, J.P. 144, 184
Verma, O.P. 114
Verner, J. 245
Victoria, J.K. 231
von Haartman, L. 221

Wagner, S.S. 246
Walankiewicz, W. 230
Waldman, B. 129
Wallace, A.R. 224–5
Walters, D. 121
Walters, S.M. 15
Wanscher, J.H. 220
Weatherhead, P.J. 121, 237, 246–7
Webber, M.I. 196
Wesolowski, T. 230, 246
Westman, B. 100
Westneat, D.F. 120–1, 210, 237
Wetton, J.H. 121
White, B.N. 121, 237
White, G. 218, 230
Whittingham, L.A. 184, 210, 239
Wiley, R.H. 101–2
Williams, G.C. 7, 169
Williamson, K. 230
Willson, M.F. 245
Wilson, V. 120
Wingfield, J.C. 195
Winkler, D.W. 171
Wittenberger, J.F. 245–6
Wolf, L. 236
Wood, J.D. 11
Woolfenden, G.E. 4, 103
Wooller, R.D. 78
Wrege, P.H. 120, 193, 239
Wright, J. 170, 173
Wyllie, I. 217, 224

Yamagishi, S. 231
Yasukawa, K. 246
Yom Tov, Y. 236

Zack, S.W. 121
Zann, R. 121, 237

Subject Index

Aardvark approach 6
Acanthis cannabinna, see Linnet
Acanthis flammea, see Redpoll
Acanthiza, see Thornbills
Accentor, alpine 10, 12, 114, 241
Accentor, black-throated 13
Accentor, brown 13
Accentor, Himalayan 12
Accentor, Japanese 12, 241
Accentor, Kozlov's 13
Accentor, maroon-backed 13
Accentor, Radde's 13
Accentor, robin 13
Accentor, rufous-breasted 13
Accentor, Siberian 13
Accentor, Yemen 13
Accipiter nisus, see Sparrowhawk
Acrocephalus scirpaceus, see Warbler, reed
Aegithalos caudatus, see Tit, long-tailed
Age
    mating success 76–81
    senescence 76, 78–9
    territory defence 76–85
    wing length 74
Agelaius phoeniceus, see Blackbird,
        red-winged
Alexander, H.G. 11
Anthus pratensis, see Pipit, meadow
Aphelocoma coerulescens, see Jay, scrub
Apis mellifera, see Honeybee
Aristotle 215

Babbler, grey-crowned 170
Blackbird, *Turdus merula* 17, 34, 79, 103,
        225, 229
Blackbird, yellow-headed 246
Blackbird, red-winged 82, 237
Blackcap 17
Botanic Garden, Cambridge 15–17
Botanic Garden, Oxford 42–3
Breeding cycle 87–8, 155–9
British Trust for Ornithology 29, 216, 218

Brood division 127, 177–8
Brood parasitism
    interspecific 129, 214–34
    intraspecific 124
Bullfinch 17, 228–9
Bunting, reed 229, 232
Bunting, snow 170
Buteo galapagoensis, see Hawk, Galapagos

Carduelis carduelis, see Goldfinch
Carduelis chloris, see Greenfinch
Cats 17, 33, 137, 186
Certhia familiaris, see Tree-creeper
Ceryle rudis, see Kingfisher, pied
Chaffinch 17, 228–9, 232
Chick feeding
    frequency 132–5, 171–6
    relation to mating success 117–9, 196–201,
        205–8
    stable efforts 169–171, 192
Chick recognition 126–130
Chick survival 134–8, 175, 180–4
Chick weight 134–138
Chiroxiphia linearis, see Manakin, long-tailed
Chrysococcyx lucidus, see Cuckoo, shining
        bronze
Clamator glandarius, see Cuckoo, great
        spotted
Cloaca pecking 106–9
Cloaca protuberance 109–12
Clutch size 6, 88, 149–51, 244–5
Colour-ringing 17–19
Copulations
    alpha versus beta male success 88–96, 162
    and chick feeding 117–9, 196–201,
        205–210
    and clutch size 149–51
    female encouragement 95, 162
    and infanticide 140–4, 200–1
    and paternity 119–26, 162–5, 196–9,
        202–5
    in polygynandry 184–9

onset 89
rate 103–6
valuation by males 201–2
Corvus corone, *see* Crow, carrion
Corvus monedula, *see* Jackdaw
Coturnix coturnix japonica, *see* Quail,
    Japanese
Cowbird, bay-winged 233
Cowbird, brown-headed 221, 227
Cowbird, giant 227
Cowbird, screaming 233
Crow, carrion 17, 137, 187
Cuckoo, common 4, 128, 214–34
Cuckoo, great spotted 231
Cuckoo, shining bronze 226
Cuculus canorus, *see* Cuckoo, common
Cyanopica cyana, *see* Magpie, azure-winged

Darwin, C. 15, 249
Darwin's finches 4, 7
Deer, red 7
Desertion, of mate 5, 170–1, 236–7
Dispersal 29–31, 39–41
Distribution 10–15
DNA fingerprinting 120–8, 162, 191, 197–8,
    202–5, 208
Dog, African wild 240
Dominance
    alpha and beta males 24–7
    and chick feeding 172–6, 196–8, 205–7
    effect of removals 81–5
    in polygynandry 184–192
    winter feeding sites 34–7, 39
Dryocopus martius, *see* Woodpecker, black
Ducks 105

Edward Grey Institute 4
Egg laying 88, 96–8
Emberiza schoeniculus, *see* Bunting, reed
Erithacus rubecula, *see* Robin
Estrildids 233
Evolutionary arms race 227, 231
Evolutionary lag 228–32
Evolutionarily stable strategy 170, 192–3

Falco subbuteo, *see* Hobby
Falco tinnunculus, *see* Kestrel
Fertile period 82–3, 88, 96–9, 200–5
Ficedula albicollis, *see* Flycatcher, collared
Ficedula hypoleuca, *see* Flycatcher, pied
Field experiments 19–21
Fights
    between females 140–4
    between males 81–5, 206–7

Finch, Bengalese 111
Finch, zebra 111, 204
Finland
    dunnock mating systems 42–3
    population increase 11–14
Fish, breeding systems 5
    parental effort 169
Flycatcher, collared 4
Flycatcher, pied 170, 228–9, 246–8
Flycatcher, spotted 17, 228–9
Food
    and brood division 177–8
    diet, adult 14–16
    diet, chicks 132
    effect on mating system 58–67, 238
    effect on territory size 58–63
    effect on time budgets 60–61, 66–8
Foraging
    costs of mate guarding 91–3
    diet 14–16
    dominance in winter 34–6
Forest clearance 11, 14, 230–1
Fringilla coelebs, *see* Chaffinch
Fringillidae 228

Gallinula chloropus, *see* Moorhen
Garrulus glandarius, *see* Jay
Goldcrest 17
Goldfinch 17
Greenfinch 17, 228–9
Ground squirrel, Belding's 129
Gulls 4

Hawk, Galapagos 154, 242
Henslow, J.S. 15, 17
Hirundo rustica, *see* Swallow, barn
Hobby 17
Honeybee 129
Hormones, and parental care 195

Iceland, no cuckoos 229–30
Incubation 88, 96–8, 117
Indirect fitness 242
Individual differences in behaviour 6–7
Infanticide 140–4, 200–1

Jackdaw 15
Jay, *Garrulus glandarius* 17
Jay, scrub 4, 103

Kestrel 17, 33, 34, 177
Kin recognition 128–30

King Lear 230
Kingfisher, pied 239

Langurs 143
Linnet 17, 228–9
Lion 143, 241
Lonchura striata, see Finch, Bengalese
Lycaon pictus, see Dog, African wild

Magpie, azure-winged 231
Magpie, black-billed 88, 231
Mammals, breeding systems 5
  infanticide 142–3
  parental effort 169
  paternity measurement 120
Manakin, long-tailed 19
Martin, sand; *see* Swallow, bank
Mate guarding
  behaviour 88–91
  costs 91–3
  duration 96–8
  influence on copulations 93–6, 210–2
  influence of territory characteristics 93–4
  intensity 91–6
  polygynandry 184–8
  removal experiments 99–100
  timing 91–2
Maternity
  measurement 120–6
Melanerpes formicivorus, see Woodpecker,
    acorn
Model egg experiments
  copulation value 201–2
  cuckoo egg mimicry 218–24
Molothrus ater, see Cowbird, brown-headed
Molothrus badius, see Cowbird, bay-winged
Molothrus rufoaxillaris, see Cowbird,
    screaming
Monogamy
  chick feeding 118, 133–4, 171–6, 207–10
  evolution of 235–7
  fledgling care 177–8
  formation 26, 48–50
  male and female territories 25–6, 46–8
  male removal 82–3
  mating and paternity 121–4
  reproductive success 141–8, 161–8
Moorhen 170
Mortality
  adult 32–7
  chicks 134–8, 175, 180–2
  effect on lifetime success 155, 169
  effect on mating system 36–7, 53–6
Motacilla alba, see Wagtail, pied and white
Muscicapa striata, see Flycatcher, spotted
Muscicapidae 228

Nest
  structure 87
  type and access to cuckoos 228
New Zealand, dunnock introduced 14
Nun 120

Owl, tawny 17, 33–4

Panthera leo, see Lion
Parus ater, see Tit, coal
Parus caeruleus, see Tit, blue
Parus major, see Tit, great
Passer domesticus, see Sparrow, house
Paternity
  markers 128–30
  and mating success 123–6, 161–6, 190–3,
    202–5
  measurement 119–23
  and parental care 126–8, 190–3
Phenotype matching 129
Phoenicurus phoenicurus, see Redstart
Phylloscopus trochilus, see Warbler, willow
Phylloscopus warblers 246
Pica pica, see Magpie, black-billed
Pipit, meadow 216–220, 222–3, 229–30
Plectrophenax nivalis, see Bunting, snow
Ploceidae, *see* Weaverbirds
Polyandry
  chick-feeding 117–9, 133–5, 171–6,
    196–201, 205–10
  competition for matings 88–91, 103–6,
    210–2
  evolution of 237–42
  fledgling care 177–8
  formation 50–51
  male and female territories 24–27, 46–51
  males not relatives 39–40
  mating and paternity 120–4, 162
  removal experiments 67–71, 99–100, 137,
    146–8, 199–201
  reproductive success 141–6, 161–5
  Tibetan 241
Polygynandry
  chick-feeding 118–9, 133–5, 184–93
  egg desertion 140–4
  females not relatives 40–41
  fledgling care 177–8
  formation 51–3
  males not relatives 39–40
  male and female territories 24–28, 46–51
  mate guarding 184–9
  mating and paternity 120–4
  removal experiments 67–71, 199–201
  reproductive success 141–6, 165–6

Polygyny
  chick-feeding 133–6, 176
  evolution of 236–7
  females not relatives 40
  formation 49–50
  male and female territories 26–7, 47–8
  male help 181–4
  reproductive success 133, 141–3, 161–5
Polygyny threshold model 245–8
Pomatostomus temporalis, *see* Babbler,
    grey-crowned
Population structure 28–33
Porphyrio porphyrio, *see* Pukeko
Poultry 112–3
Primeval woodland 11, 230
Protein polymorphism 120
Protonotaria citrea, *see* Warbler,
    prothonotary
Prunella atrogularis, *see* Accentor,
    black-throated
Prunella collaris, *see* Accentor, alpine
Prunella fagani, *see* Accentor, Yemen
Prunella fulvescens, *see* Accentor, brown
Prunella himalayana, *see* Accentor,
    Himalayan
Prunella immaculata, *see* Accentor,
    maroon-backed
Prunella koslowi, *see* Accentor, Kozlov's
Prunella montanella, *see* Accentor, Siberian
Prunella ocularis, *see* Accentor, Radde's
Prunella rubeculoides, *see* Accentor, robin
Prunella rubida, *see* Accentor, Japanese
Prunella strophiata, *see* Accentor,
    rufous-breasted
Prunellidae 10–13
Pukeko 69, 242
Pyrrhula pyrrhula, *see* Bullfinch

Quail, Japanese 129, 225

Rabbit 58
Redpoll 17
Redstart 221, 229
Regulus regulus, *see* Goldcrest
Relatedness
  between alpha and beta males 39–40
  between females 40–1
  and chick feeding 117–9, 242
  and cooperation 37–8, 242
Removal experiments, effects on
  clutch size 151
  male status 81–5
  mating system 67–71
  matings 99–100

parental effort 173–6, 199–202, 205–10
paternity 202–5
reproductive success 137, 146–8, 236
valuation of copulations 201–2
Reproductive success
  lifetime success 154–5
  number young fledged 138–48
  seasonal success 155–60
Riparia riparia, *see* Swallow, bank
Robin, *Erithacus rubecula* 6, 17, 177,
    216–22, 229, 237–8
Rodents 128, 143, 225

Saguinus fuscicollis, *see* Tamarin,
    saddle-backed
Scaphidura oryzivora, *see* Cowbird, giant
Sciurus carolinensis, *see* Squirrel, grey
Seabirds 105, 236
Selection thinking 7
Senescence 76–79
Sex ratio
  breeding population 36–8
  mating system 36–8, 41–3
  winter severity 34–5
Size
  and mating system 74, 77
  tarsus length 74–7
  wing length 74–7
Shorebirds 236
Song
  effect of food 68
  female response 100–3
  individual variation 100–3
  mating system formation 50
Sparrowhawk 17, 33
Sparrow, house 17, 34, 144
Sparrow, song 6
Species-specific behaviour 6
Sperm
  ejection by females 107–10, 204
  numbers 111
  storage in females 97, 109–13, 204
Spermophilus beldingi, *see* Ground squirrel,
    Belding's
Squirrel, grey
  as nest predator 17, 137, 225
  hair for nest lining 87
Starling 17, 170, 173, 229
Strix aluco, *see* Owl, tawny
Sturnus vulgaris, *see* Starling
Survival, *see* Mortality
Swallow, barn *Hirundo rustica* 143, 229
Swallow, bank *Riparia riparia* 88, 97, 128
Swallow, tree *Tachycineta bicolor* 144
Sylvia atricapilla, *see* Blackcap
Sylvia curruca, *see* Whitethroat, lesser

Tachycineta bicolor, *see* Swallow, tree
Taeniopygia guttata, *see* Finch, zebra
Tamarin, saddle-backed 239–40
Tasmanian native hen 37
Territory
  boundary displays 24, 49
  characteristics and mate guarding 93–4
  coalescence 50–3
  male and female overlap 24–28, 162
  maps 25, 27–8, 48, 51–2, 167
  quality and reproductive success 146–8
  removal experiments 67–71, 81–5, 99–100
  size and age 79–81
  size and food distribution 58–63, 162
  size and mating system 46–8, 56–71,
    237–8
  vacancies 53–8
  value to males 81–85
Testis, size 111
Thornbills 226
Thrush, song 17, 225, 229
Tibetan polyandry 241
Time budgets 61, 68
Tit, blue 17, 170, 229
Tit, coal 17
Tit, great 4, 17, 82, 170, 229

Tit, long-tailed 17
Tree-creeper 17
Tribonyx mortierii, *see* Tasmanian native hen
Troglodytes troglodytes, *see* Wren
Turdus merula, *see* Blackbird
Turdus philomelos, *see* Thrush, song

Viduines 233

Wagtail, pied 216–22, 229
Wagtail, white 229–30
Warbler, prothonotary 227
Warbler, reed 216–22, 225, 229, 232
Warbler, sedge 229
Warbler, willow 17, 81
Weaverbirds 236
Wheatear 229
Whitethroat, lesser 17
Woodpecker, acorn 4, 38, 144, 154, 242
Woodpecker, black 11
Wren 17, 229

Xanthocephalus xanthocephalus, *see*
  Blackbird, yellow-headed